AF793567

Advances in Soil Science

Advances in Soil Science

B.A. Stewart, Editor

Editorial Board

R. Lal
C.W. Rose
U. Schwertmann
B.A. Stewart
P.B. Tinker
R.J. Wagenet
B. Yaron

Advances in Soil Science

Volume 12

Edited by B.A. Stewart

With Contributions by
S.K. Jalota, B.D. Kay, S. Komarneni, P.B. Malla,
E. Murad, S.S. Prihar, and M.E. Sumner

With 49 Illustrations

Springer-Verlag
New York Berlin Heidelberg
London Paris Tokyo Hong Kong

B.A. Stewart
USDA Conservation & Production Research Laboratory
Bushland, Texas 79012, U.S.A.

ISSN: 0176-9340

Printed on acid-free paper.

© 1990 by Springer-Verlag New York Inc.
Softcover reprint of the hardcover 1st edition 1990

All rights reserved. This work may not be translated or copied in whole or in part without the written permission of the publisher (Springer-Verlag New York, Inc., 175 Fifth Avenue, New York, NY 10010, U.S.A.), except for brief excerpts in connection with reviews or scholarly analysis. Use in connection with any form of information storage and retrieval, electronic adaptation, computer software, or by similar or dissimilar methodology now known or hereafter developed is forbidden. The use of general descriptive names, trade names, trademarks, etc., in this publication, even if the former are not especially identified, is not to be taken as a sign that such names, as understood by the Trade Marks and Merchandise Marks Act, may accordingly be used freely by anyone.

Typeset by Kachina Typesetting, Inc., Tempe, Arizona.

9 8 7 6 5 4 3 2 1

ISBN-13:978-1-4612-7964-8 e-ISBN-13:978-1-4612-3316-9
DOI:10.1007/978-1-4612-3316-9

Preface

Soil is formed from the physical and chemical weathering of rocks—processes described historically because they involve eons of time—by glaciation and by wind and water transport of soil materials, which are later deposited in deltas and loessial planes. Soil undergoes further transformations over time and provides a habitat for biological life and a base for the development of civilizations. Soil is dynamic—always changing as a result of the forces of nature and particularly by the influences of man. Soil has been studied as long as history has been documented. Numerous references to soil are found in historical writings such as Aristotle (384–322 BC), Theophrastus (372–286 BC), Cato the Elder (234–149 BC), and Varro (116–27 BC). Some of the earliest historical references have to do with the erosional forces of wind and water.

The study of soils today has taken on increased importance because a rapidly expanding population is placing demands never before experienced on the soil. This has led to an increase in land degradation. Land degradation is one of the most severe problems facing mankind. Volume 11 of *Advances in Soil Science* was devoted entirely to this critical area of soil science.

This series, *Advances in Soil Science,* was established to provide a forum for leading scientists to analyze and summarize the available scientific information on a subject, assessing its importance and identifying additional research needs. This goal seems even more appropriate today than in 1982 when the idea of the series was formulated. Much has been learned about our soil resources. The principles learned and the technology developed need to be used to increase food production and sustain the productivity of the resource base. *Advances in Soil Science* fills a gap between the scientific journals and the comprehensive reference books. Scientists can delve in depth on a particular subject relating to soil science. Contributors are asked in particular to develop and identify principles that have practical applications to both developing and developed agricultures.

Advances in Soil Science was formulated to be international in scope and to cover all subjects relating to soil science. This volume continues in that format in that it contains contributions from scientists from Canada, India, the Federal Republic of Germany, and the United States on subjects ranging from very basic research on nitrogen fixation and the study of clay mineralogy to field studies concerned with soil structure and water evaporation. Although we consider our audience to be primarily scientists and students of soil science, the series provides technical information to anyone interested in our natural resources and man's influence on these resources. Research in the future will focus on systems that are resource efficient and environmentally sound. The need to optimize crop production while conserving the resource base has never been greater.

The quick acceptance of *Advances in Soil Science* by both authors and readers has been very gratifying and confirms our perception that a need did exist for a medium to publish soil science reviews. I want to thank the authors for their excellent contributions and cooperation. I also want to thank the members of the Editorial Board for their assistance in selecting such competent authors and the Springer-Verlag staff for their kind assistance and counsel. Lastly, and most importantly, I want to thank the readers for their acceptance and use of *Advances in Soil Science*.

B.A. Stewart
December, 1989

Contributors

S.K. Jalota, Department of Soils, Punjab Agricultural University, Ludhiana 141004, India.

B.D. Kay, Department of Land Resource Science, University of Guelph, Guelph, Ontario N1G 2WI, Canada.

S. Komarneni, Materials Research Laboratory and the Department of Agronomy, Pennsylvania State University, University Park, Pennsylvania 16802, U.S.A.

P.B. Malla, Materials Research Laboratory, Pennsylvania State University, University Park, Pennsylvania 16802, U.S.A.

E. Murad, Lehrstuhl für Bodenkunde, Technische Universität München, D-8050 Freising-Weihenstephan, Federal Republic of Germany.

S.S. Prihar, Department of Soils, Punjab Agricultural University, Ludhiana 141004, India.

M.E. Sumner, Department of Agronomy, University of Georgia, Athens, Georgia 30602, U.S.A.

Rates of Change of Soil Structure Under Different Cropping Systems

B.D. Kay*

I. Introduction

The structure of soil can determine both the effectiveness and the impact of farming practices. Soil structure influences the extent to which crop breeding and the management of weeds, insects, disease, soil fertility, and water are manifested in increased crop yields. Soil structure also influences the loss of agricultural chemicals through erosion and leaching and can, therefore, have a significant bearing on the environmental impact of some agricultural practices.

Soil structure is very sensitive to human activities. The extent of change

*Department of Land Resource Science, University of Guelph, Guelph, Ontario, N1G 2WI, Canada.

arising from these activities can be considered over time scales ranging from hours to centuries. Agricultural practices continue to evolve, and the introduction of each new management practice can alter the rate of change of soil structure. The rate of change in soil structure occurring within a time scale compatible with a farmer's planning horizon is of paramount importance. For most farmers this time scale is one of years rather than days or decades, and it is within this time scale that choices are made with regard to production practices which determine the rate of change in soil structure.

This paper reviews the current understanding of the development of soil structure and its rate of change under different agricultural practices. Emphasis is placed primarily on soils in temperate regions of the world and the rates of change in structure under cropping systems in time scales of years. The paper concludes with a discussion of conceptual models that are being developed to characterize the rate of change of soil structure.

II. Soil Structure

A. What Is It?

Soil structure can be considered at dimensions ranging from 10^{-7} to 10^{0} m. Figure 1 relates this range of dimensions to the nature of solid materials, pores, and organisms that may be encountered in soil. Soil structure can also be considered in terms of the spatial heterogeneity (Dexter, 1988) and the temporal variability of the different components or properties of soil. For the purpose of this chapter soil structure will be described in terms of form, stability, and resiliency.

The term *structural form* will be used to describe the heterogeneous arrangement of solid and void space that exists in soil at a given time. Structural form refers to total porosity, pore size distribution, and continuity of the pore system. Structural form also refers to the arrangement of primary soil particles into hierarchical structural states identified on the basis of failure zones of different strengths.

Failure zones occur where the volume fraction of primary soil particles that are strongly bonded to one another is low. Failure zones may have a low volume fraction of particles (i.e., the sites of cracks or a high volume fraction of pores) or occur where soil particles are weakly bonded to one another. The application of stress to a clod of soil will result in fracture along failure zones, giving rise to macroaggregates (see Figure 1). The structural units which are released on application of increasing stress reflect the hierarchical nature of soil structure: with application of increasing stress the macroaggregates can be broken down progressively into microaggregates, domains, quasicrystals, and ultimately individual primary soil particles. This aspect of structural form is an extension of the definition of aggregates as units in which primary particles are held together by ties stronger than the ties between adjacent aggregates (Nikiforoff, 1941).

The *stability* of a soil's structure is the ability of the soil to retain its ar-

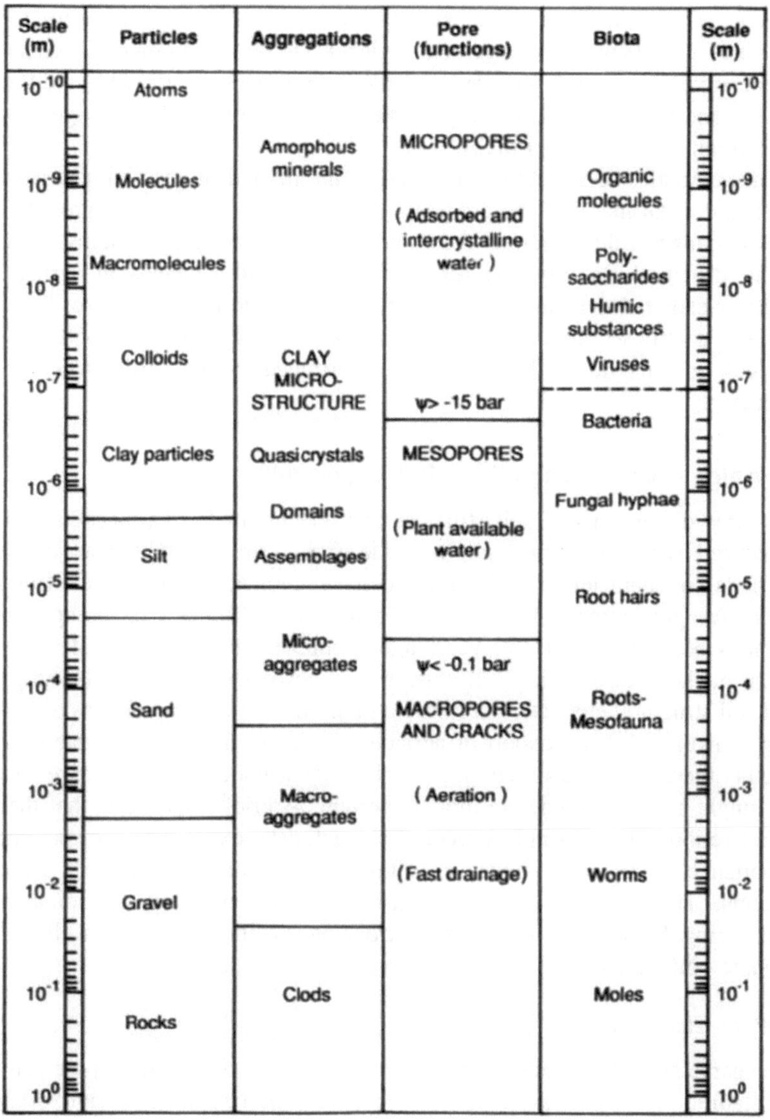

Figure 1. Variation in size of different components of soil. (J. M. Oades, personal communication.)

rangement of solid and void space when exposed to different stresses. Stability characteristics are generally specific for a characteristic of structural form and the type of stress being applied; for example, the resistance of a pore system to compressive stresses will be different from the resistance of clay domains to dispersion by osmotic stresses.

The term *resiliency* is used to describe the ability of a soil to recover its structural form through natural processes when the applied stresses are reduced or removed. Although this term has not been used specifically in relation to soil structure, it offers the advantage of a single term to refer to a range of recovery or regenerative processes. Included among such processes are tilth mellowing (Utomo and Dexter, 1981a), self-mulching (Blackmore, 1981), and age hardening (Utomo and Dexter, 1981b).

Characteristics of structural form determine the hydrologic characteristics of soil and control many of the processes that are important to plant growth. The rate of change of the structural form under different cropping systems is related to the structural stability and resiliency of the soil. Changes in agricultural practices can lead to changes in structural form by altering the stresses to which the soil is exposed or its stability. The extent to which the structural form changes on removal of the stresses is determined by the soil's structural resiliency.

B. Which Characteristics Are Most Important?

The three dimensions of soil structure and the multitude of processes that are influenced by soil structure result in a wide array of soil structural characteristics. Methods to characterize soil structure have evolved through the efforts of successive generations of researchers, with one of the earliest compendia of methods being prepared 150 years ago (Schubler, 1840). Some of the more recent methods to describe soil structure are reviewed by Dexter (1988).

If soil structural characteristics have a major impact on processes as diverse as crop growth and the transport of agricultural chemicals, it is reasonable to ask whether there are some structural characteristics which, among the large number that might be considered, are most important. A brief response is that there is little likelihood that a single structural characteristic exists that is of paramount importance in all soils, under all climatic conditions, and for all cropping systems. There are, however, characteristics that are very important when considering the production of specific crops on specific soils under well-defined climatic conditions.

1. Pore Characteristics

The characteristics of macropores (porosity, pore size distribution, pore continuity, stability of the pore system, resiliency) are important in infiltration, drainage and aeration, and leaching of agricultural chemicals. The importance of these structural characteristics has led Gibbs and Reid (1988) to equate soil structure with pores greater than 100 μm in diameter and begin to develop a model that describes changes in the proportion of such pores under different cropping systems. This characteristic of structure will be particularly important where soil, climate, or cropping systems make it desirable to increase or, conversely, decrease the extent of drainage or leaching in order to increase yields or diminish the extent of erosion or leaching. Such pores are also important in the development of roots in compacted soils (Dexter, 1986a, 1988b; Stypa et al., 1987).

The porosity, which retains water available for plant growth, and its associated characteristics (pore size distribution, pore continuity, stability, resiliency) are of crucial importance in rain-fed agriculture where water may become limiting during critical stages of crop growth. A detailed analysis of long-term yields and fertilizer response of different crops at Rothamsted led Johnston (1986) to conclude that yields of shallow rooted crops such as potatoes *(Solanum tuberosum)*, spring wheat *(Triticum aestivum)*, sugar beet *(Beta vulgaris)*, and spring barley *(Hordeum vulgare)* can be increased significantly (up to 36.6% for potatoes) by increases in organic matter content (from 0.76 to 2.03% organic carbon for potato trials) and that the increased yields are due to increased amounts of water available for plant growth. Yield increases were significantly less for fall-planted crops, which were deeper rooted and could therefore benefit from water held at greater depth in the profile. Infiltration and drainage presumably did not present a limitation to crop growth on any of the soils investigated.

Under circumstances where both drainage and available water are important, Thomasson (1978) has used the relative proportion of macropores (which he defined as pores greater than 60 μm in diameter) and mesopores (pores 60 to 0.2 μm in diameter) to define four classes of soil structure. The best class has a macroporosity $\geq$15% and a mesoporosity of 20 to 35%. The worst class has a macroporosity of <5% and a mesoporosity of <35%.

2. Strength Characteristics

Soil strength can impose limitations on plant yields if it results in reduced root growth under circumstances in which water and nutrient supply become limiting. Soil strength is strongly dependent on moisture content and increases as soils become drier. Letey (1985) has defined a nonlimiting water range (Figure 2) in which the upper end of the range is controlled by macroporosity and its relation to aeration, and the lower end of the range can be determined by the water content at which either strength or available water limits plant growth. Limitations by strength or available water are incorporated into the model developed by Dexter (1987a) to describe root elongation. The relative rate of elongation of the root tip R/R_{max} is described by an equation of the form

$$\frac{R}{R_{max}} = - \frac{\Psi_0}{\Psi_w} + e^{-0.6931(Q_p/Q_{0.5})} \qquad (1)$$

where Ψ_0 and Ψ_w are the soil water potential and the wilting point potential of the plant species, respectively, Q_p is the resistance to penetration measured by a penetrometer, and $Q_{0.5}$ is the value of the penetrometer resistance that reduces relative root elongation to 0.5 for the plant species. Published values of $Q_{0.5}$ for different species are summarized in Dexter (1987a) and range from 0.72 to 2.03 MPa. The relations between R/R_{max}, Ψ_0, and Q_p are shown in Figure 3.

3. Stability of Surface Soils

In those circumstances in which it is necessary to create a seedbed, stability characteristics such as friability (Utomo and Dexter, 1981c) determine the

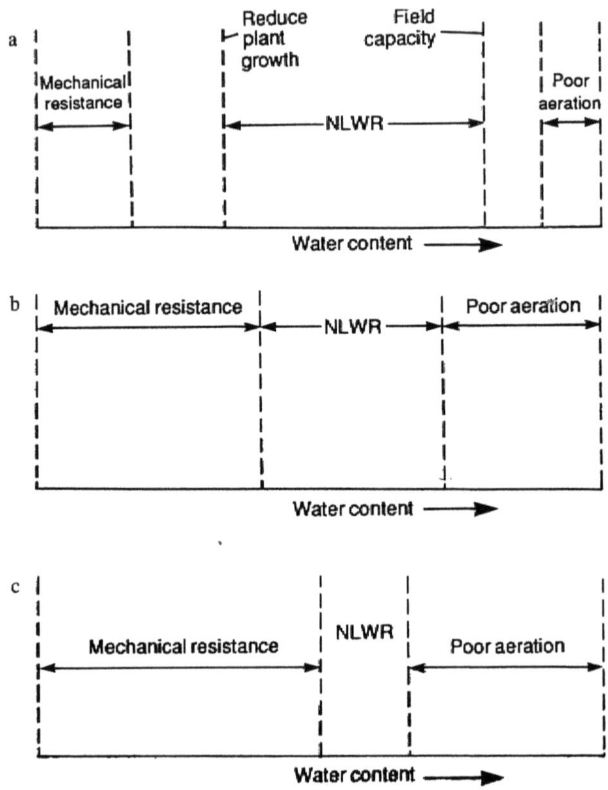

Figure 2. Generalized relationships between soil water content and factors restricting plant growth in a soil. The limiting factors progressively increase in magnitude, resulting in a decrease in the nonlimiting water range (NLWR) in conditions *A* to C. [From Letey (1985).]

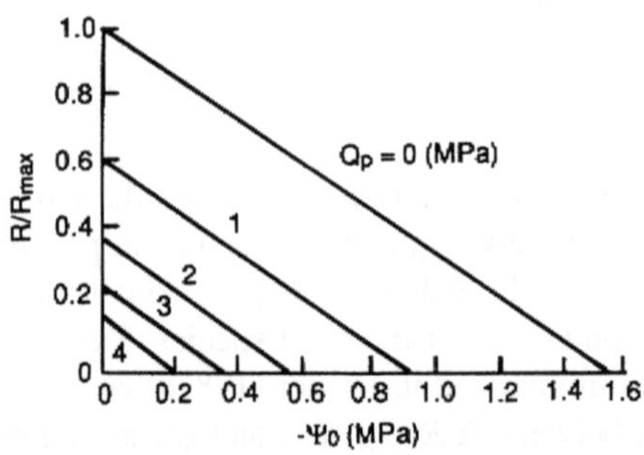

Figure 3. Values of relative rate of root elongation $R/R_{\max}$ under combinations of conditions of soil water potential, Ψ_0 and penetrometer strength Q_p, predicted from Equation (1). [From Dexter (1987a).]

aggregate size distribution created by tillage. Aggregate size distribution can, in turn, have a major influence on early shoot and root growth (Donald et al., 1987; Logsdon et al., 1987; Misra et al., 1988).

The persistence of the characteristics of the seedbed, and its susceptibility to crust formation and erosion, are related to characteristics such as wet aggregate stability and dispersible clay content (Egashira et al., 1983; Miller and Baharuddin, 1986; Abu-Sharar et al., 1987).

In addition to the direct effects considered above, indirect effects of soil structural characteristics on soil productivity also exist. For instance, crops growing in environments in which macroporosity or soil strength are limiting are also more susceptible to disease (Tu and Tan, 1988). Disease enhances the detrimental influence on crop yield of reduced uptake of nutrients and water as a consequence of reduced root growth.

The relative significance of different structural characteristics clearly varies with soil, climate, and crop. The relative importance attached to a given structural characteristic is determined by the influence of that characteristic on the soil property or process that imposes the greatest limitation on the yield of a specific crop, or the transport of sediments or agricultural chemicals, under specific climatic conditions.

III. Processes Responsible for Changing Soil Structure

A. Structural Form

The structural form existing in a soil at any given time represents the net effect of processes that are changing the structural form at different scales and at different rates. Drying, wetting, freezing, thawing, consolidation, biopore formation as well as compression, shear, and brittle fracture by roots, animals, or equipment can alter the characteristics of pores, failure zones, and aggregates in soil.

1. Drying and Wetting

Drying results in an increase in the effective stress (Croney et al., 1958) or intergranular contact pressure. In saturated soils that are not under a load or confining pressure, the effective stress is equal to the swelling pressure (Groenevelt and Kay, 1981). In unsaturated soils the effective stress is proportional to the pressure drop across the menisci of water films surrounding the particles, that is, soil water suction. The increase in effective stress on drying has the least effect on the structural form of soils in which the mineral particles are fully in contact with one another when saturated and the geometric relation of one particle to its nearest neighbors does not change as the effective stress increases. Such soils have been called incompressible (Croney and Coleman, 1954), noncohesive, or nonswelling soils. The structural form of other soils will change in response to increases in effective stress. The size scale at which the changes will occur depends on the relative proportions of noncohesive and cohesive materials. Soils with a skeleton of noncohesive materials but containing cohesive materials in interstices may exhibit little macroscopic change in structural form with increasing effective stress. However, changes on a microscale will occur within the interstitial cohesive materials. Increasing the proportion of cohesive materials and increasing the separation of the noncohesive materials will result in the whole soil behaving increasingly like its cohesive components (Towner, 1986).

The shrinkage characteristics of the cohesive materials in soils are determined by mineralogy, the nature and concentration of electrolyte in the pore fluid (Mitchell, 1976), as well as by exchange capacity and specific surface area (Low, 1980). Shrinkage of cohesive materials generates tensile stress, and cracks are initiated in the soil when the tensile stress exceeds the tensile strength. The initiation and propagation of cracks will depend on the extent and the spatial variation in water extraction and, consequently, the variation in both stress generation and tensile strength. Variations in crack formation and propagation will also depend on the spatial variation in tensile strength and the elastic properties of the soil. The tensile strength of soils increases with decreasing water content. In the case of molded clay, cracking is initiated at a unique water content, which is independent of the initial water content and can be determined from the relation between tensile strength and water content (Towner, 1987). The presence of relatively large incompressible particles such as sand grains in a molded clay matrix can induce stress in the clay matrix as shrinkage occurs and can initiate cracking of the clay matrix around single grains or between pairs of closely spaced grains (Towner, 1988). Repeated wetting and drying increases the variability in the strength of failure zones and therefore in tensile strength (Dexter et al., 1984).

The extent of crack formation and the spatial distribution of cracks are related to the magnitude of desiccation and the spatial variation in water removal from soil. An evaluation of the effect of 10 different grass species on the structural form over 10 years and the observation that the largest area of planar voids was produced where grasses resulted in the greatest yields, led Grevers and de Jong (1988) to suggest that differences in soil desiccation by grass roots were mainly responsible for differences in soil macroporosity. The location of cracks will be influenced by the pattern of water extraction in soil: cracks will develop where the tensile strength is lowest, that is, where the soil is wettest. This accounts for the observation of Fox (1964) and others that large cracks tend to form midway between the rows of crops. Dasog et al. (1988) have observed lower cracking intensity (as measured by volumes of cracks per hectare) under natural grassland than under wheat. This may be due to a more uniform pattern of moisture extraction under grass than occurs under a crop planted in rows and to the greater variability in the strength of failure zones (i.e., greater friability) often associated with soils under grass (Chan, 1989a, 1989b). The shrinkage in the grassland soil would be accommodated by the formation of a large volume of microcracks, which would not be included in the measurements made by Dasog et al. (1988).

Rewetting soils in which desiccation cracks have formed will result in swelling and a tendency for the cracks to close. Swelling may involve reorientation of particles as well as the breakage of bonds that were created between soil particles as a consequence of drying. Swelling appears to proceed rapidly in the first 24 hours after rewetting (Ladd, 1960), but can continue at greatly reduced rates for months thereafter (Emerson, 1955). The formation of cracks is coincident with a loss of porosity in the intercrack space, and therefore a persistence of cracks on rewetting implies a persistence of increased bulk density in the intercrack zone and continued changes in the pore size distribution of the whole soil.

Rewetting can also introduce cracks at a microscale as a consequence of differential swelling of the soil matrix and the pressure of entrapped air exceeding the tensile strength of the surrounding matrix (Panabokke and Quirk, 1957; Stengel, 1988). The extent of failure zone development depends on the rate of wetting, and the terms *slaking* (see Panabokke and Quirk, 1957) and *tilth mellowing* (Utomo and Dexter, 1981a) have been employed to describe differences in the extent of failure zones produced by the same processes. McKenzie and Dexter (1985) suggested mellowing can be viewed as partial slaking in the sense that soil aggregates or clods weaken but do not completely fragment into smaller pieces. Rapid wetting rates create the most extensive failure zones (slaking), slightly slower wetting rates produce microcracks or less extensive failure zones (tilth mellowing), while even slower rates do not introduce failure zones that are extensive enough to reduce tensile strength. The extent of microcrack formation must depend on both the magnitude of the stresses produced and the distribution of strength of existing failure zones. The magnitude of stresses produced on wetting increases with the rate of wetting. Therefore there must exist a critical wetting rate at which the stresses that are generated exceed the strength of enough of the weakest failure zones that the tensile strength of the soil begins to fall, that is, mellowing begins. With further increases in wetting rates a critical wetting rate may be achieved at which the soil has zero tensile strength, that is, slaking occurs. Grant and Dexter (1989), working with molded dried soils (soils with a narrow distribution of strength of failure zones), found that the critical wetting rates to initiate mellowing of two soils were 7×10^{-7} and 1×10^{-5} kg kg^{-1} s^{-1} and the critical wetting rates to initiate slaking were 4×10^{-3} and 5×10^{-5} kg kg^{-1} s^{-1}. These wetting rates are relatively rapid; the critical wetting rate for mellowing corresponds to the initial wetting rate of soil at 1.0 to 1.5 MPa water tension placed in contact with a free water source. Soils that possess a range of strengths of failure zones, some of which are much weaker than the strength of the soil used by Grant and Dexter (1989), would be expected to develop microcracks at slower rates of wetting.

Increased numbers of wetting and drying cycles result in a progressive decrease in aggregate strength (Dexter et al., 1984) and can result in a progressive decrease in aggregate size (Shiel et al., 1988). Larger aggregates would be expected to have a larger number of failure zones with low strength and would be expected to be more susceptible to wetting and drying. Shiel et al. (1988)

observed that the rate of decrease in aggregate size with the number of wetting and drying cycles declined with the aggregate size.

The distribution of strength of failure zones is also changed by tillage, with some failure zones necessarily weakening. This probably accounts for the effectiveness of wetting and drying under field conditions in enhancing the extent of failure zones in aggregates subsequent to tillage (Utomo and Dexter, 1981a) since the wetting rates must have been much smaller than the critical wetting rate for mellowing observed by Grant and Dexter (1989).

The impact of wetting and drying cycles on structural form will vary with the cohesive characteristics of a soil. For instance, experience in western Europe (Heinonen, 1982) indicates that the minimum content of medium active clay required to make wetting and drying cycles effective in loosening compacted soil is about 15 to 20%. However, the impact of wetting and drying will also vary with the initial structural state. Wetting and drying cycles may produce an overall increase or decrease in porosity depending on whether the soil was from a site that was recently compacted or remolded, respectively. The changes would move the structural form toward the form that Heinonen (1977) defined as the "normal" form for that soil, when the physical and biological environment remains stable for an extended period.

2. Freezing and Thawing

The impact of ground freezing on the structural form of soil is determined mainly by the size and the distribution of the ice crystals that are formed (Nagasawa and Umeda, 1985). Detailed discussions on the physics of ice accumulation in freezing soils are provided in Miller (1980) and Kay and Perfect (1988). Water may freeze in situ in pores or after redistribution caused by the freezing process. The in situ freezing of water-filled pores will lead to a fracturing of the soil. Freezing often involves redistribution of water due to the development of large gradients in the soil water potential. Local redistribution may occur with ice accumulating in large pores as adjacent areas become dehydrated and shrink. The process is analagous to the formation of desiccation cracks in cohesive soil materials, in that porosity is lost in some zones and gained in others.

The most dramatic impact of ground freezing occurs when significant quantities of water move from the unfrozen zone up into the frozen zone in response to freezing-induced gradients in soil water potential. The water accumulates as discrete bands or lenses of ice. The extent of ice lens formation is controlled by the availability of water to the freezing zone (McGaw, 1972) and by soil texture (Linell and Kaplar, 1959). Soils with a high water content and which have a high silt content are most susceptible to ice lens formation. The morphology of ice lenses in soils can change significantly with texture, mineralogy, and exchangeable ions (Czeratzki and Frese, 1958). Yamada (1955) has reported ice lenses in volcanic ash soils as being a collection of slender pillars. Kay et al. (1985) observed a laminar structure of ice lenses in a silt loam, but a distinctly reticular or polygonal structure in a clay loam (Figure 4). The formation of ice lenses effectively slices up the existing structural form: the thinner and more closely spaced the lenses, the more extensive the impact on the structural form. The accumulation of water in the frozen zone results in dramatic increases in porosity in the frozen zone, and the effect is greatest near the ground surface (Kay et al., 1985). However, most of this additional porosity is lost as the soils thaw, drain, and reconsolidate (Morgenstern and Nixon, 1971; Kay et al., 1985). The residual effects of ground freezing on structural form are the formation or accentuation of failure zones that existed at the location of former ice lenses, pores which did not completely collapse, and consolidated areas in which shrinkage occurred during freezing and in which swelling occurs slowly.

The significance of the residual failure zones and any increased porosity will depend on the extent of reconsolidation, the "fit" between the two surfaces created by the ice lens, and the ability of the soil to reform bonds broken by ice lens formation. For instance, Bisal and Nielsen (1964) suggested that the impact of freezing on aggregate size distribution (a reflection of the spatial distribution of failure zones) depends, in part, on the manner of drying. Their data indicated freeze-dried soils (such as might be encountered on snow-free soil on the Great Plains of the United States), which experience no reconsolidation and have little opportunity for reforming bonds at the sites of failure zones created by feeezing, will have a higher proportion of small aggregates than soils that can reconsolidate

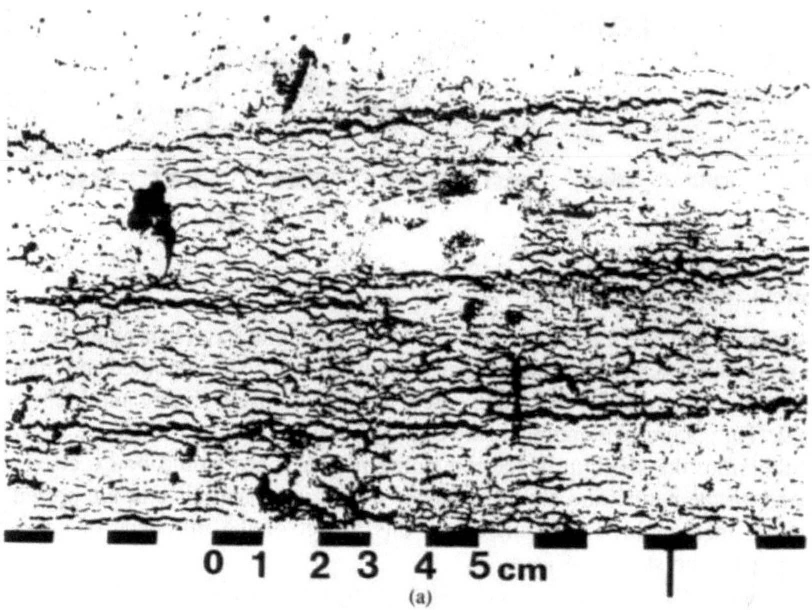

(a)

Figure 4. Macromorphology of ice-enriched soil. (a) Silt loam. (b) Clay loam. [Kay et al. (1985). Reproduced from *Soil Science Society of America Journal,* volume 49, no. 4, July–August, 1985, pp. 973–978 by permission of the Soil Science Society of America, Inc.]

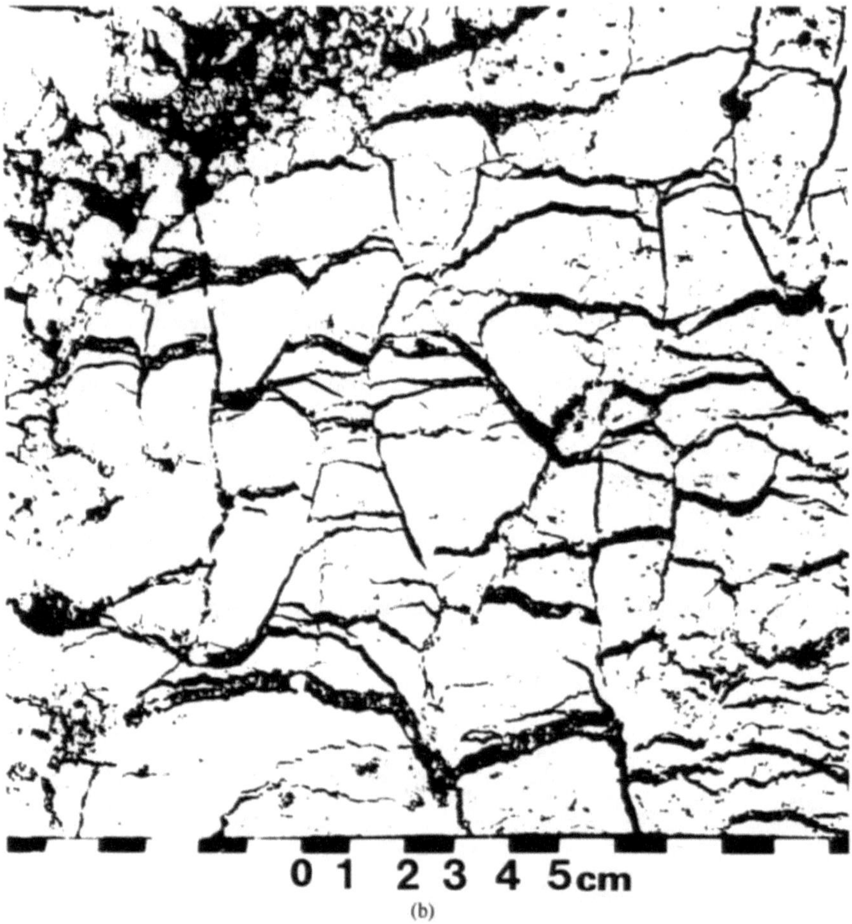

Figure 4. *Continued*

on draining and are then air-dried. Similarly, soils that have had extensive ice accumulation and have thawed but not yet drained will be in a puddled condition, and this may account for their high susceptibility to erosion (Kirby and Mehuys, 1987). The nature of bonding, which develops at the site of former ice lenses, and therefore the strength of the remaining failure zones would be expected to depend on the nature and effectiveness of stabilizing materials that are present, although there is little information on this aspect of the influence of ground freezing on the structural form.

The influence of cyclic freezing and thawing can also be detected on a microscale. Pawluk (1988) has shown that repeated freezing and thawing of cores of glacial till clay loam results in the formation of granic and metafragmic microfabrics, whereas well-sorted lacustrine sediments tend to form banded fabrics. A distinctive type of banded fabric has also been observed in sandy seasonally frozen soils (Mermut and St. Arnaud, 1981).

The significance of the changes in structural form caused by freezing depends, as for wetting and drying, on the initial structural characteristics of the soil. For instance, Chamberlain and Gow (1979) found that freezing molded reconsolidated fine-textured soils with access to a free water supply resulted in an overall reduction in porosity but an increase in vertical permeability. The largest increase in permeability occurred for the soil of highest plasticity. Benoit (1973), on the other hand, working with columns of aggregates of different size from a silt loam found that freezing and thawing caused an increasing breakdown of aggregates with increasing aggregate size and concomitant decreases in hydraulic conductivity in the columns of aggregates.

3. Deformation by Shearing, Compression, and Tensile Stresses

Shearing, compressive, and tensile stresses can be exerted during tillage and by field traffic, as well as by earthworms, plant roots, and livestock. In the simplest sense, pure shear deformation involves a change in shape at constant volume of a unit volume of soil, whereas a pure compressive strain involves a change in volume at constant shape. Deformation of soil materials in response to a uniaxial stress reflects changes in both shape and volume and is therefore a mixture of both pure shear and pure compression. Tensile strain can be considered to be the same as axial strain, but in the opposite direction.

Tillage may be carried out to control weeds, incorporate material into soil, conserve water, or improve the environment for plant growth. Soil will deform or fail in response to tillage by brittle failure, compressive failure, or tensile failure. Brittle failure occurs when a soil is compressed and the soil deforms along a few well-defined planes while little soil disturbance occurs within the soil mass between the failure planes. Compressive failure also occurs as a consequence of compressive loading, but in this case soil fails along a large number of planes and is compacted. Tensile failure arises from tensile stress; the soil density is decreased as fissures and cracks are opened up, but the soil between the failure surfaces is largely undisturbed. Tensile and brittle failure are most likely to occur when confining stress (the resistance applied by the surrounding soil to the deforming soil mass) is low, and compressive failure will occur when confining stresses are high. The higher the soil moisture content or the lower its density, the lower the confining stress must be before brittle or tensile failure can occur (Spoor, 1988). Tillage can increase the connectivity of failure zones of low strength or break bonds, thereby weakening failure zones. However, in those regions that become compacted as a consequence of tillage, failure zones may be strengthened. The nature of deformation by tillage will determine the characteristics of failure zones and the resulting aggregate size distribution, porosity, pore size distribution, and pore continuity, as well as the variation in these characteristics of structural form with depth. When tillage leads to an increase in the extent of failure zones, reduced stability is also observed (Utomo and Dexter, 1982). The propagation through soil of different stresses arising from tillage equipment and the resulting deformation of the soil matrix are reviewed in greater detail by Koolen and Kuipers (1983) and Hettiaratchi (1988).

Tillage can also enhance the impact of other processes leading to a change in structural form. The amplitude of wetting and drying cycles of surface soil is increased (Utomo and Dexter, 1981a) subsequent to tillage, with concomitant influence on shrinkage and the development of failure zones on rewetting. Tillage can also have a profound influence on the rate of loss of structure-stabilizing materials (discussed later). The persistence of changes in structural form caused by tillage is not well documented, but is obviously related to stability, climatic factors, and traffic (see Cassel, 1983; Groenevelt et al., 1984).

Increased field traffic (frequency of passes or increased machinery weight) characterizes many of the changes in today's mechanized agriculture. Hakansson et al. (1988) have suggested that the loaded wheel has become the most frequently used "tillage tool." Traffic on soils with a low ability to support a load will create deep tracks or ruts, causing extensive shear strain and resulting in smearing and the creation of a large number of failure surfaces. Soil with greater strength will decrease in volume as a consequence of both compression and shear. The extent of deformation and the depth in the soil to which it occurs depend on vehicle characteristics such as axle load and tire type (Taylor and Burt, 1987), number of passes (Gameda et al., 1987), and vehicle speed (Horn et al., 1988), as well as on soil characteristics such as moisture content (Larson et al., 1980; Tisdall and Adem, 1986; Gameda et al., 1987), texture (Larson et al., 1980), structural characteristics (Angers et al., 1987), and time since the soil was last disturbed (Utomo and Dexter, 1981b). The depth to which these structural changes occur may extend well below the normal tillage zone. Voorhees et al. (1978) have shown that the wheel traffic of normal farming operations in the northern corn belt of the United States can compact soil to a depth of 45 cm.

Changes in structural form that are caused by traffic include changes in both porosity and pore size distribution as well as the nature of failure planes. As the applied stress increases, the largest pores collapse first (Katou et al., 1987), and aggregates move closer together as the points of contact between aggregates flatten (Braunack et al., 1979). Zones of failure arising from rapid wetting or from ice crystal formation may be strengthened as the surfaces are forced closer together and subsequent opportunities for bonding at particle–particle contact points are enhanced. Biopores may be less susceptible to collapse by traffic than is generally the case for macropores with a more irregular geometry. Dexter (personal communication) has shown theoretically that vertical biopores will be unaffected by the application of an axial load. Additional details on the impact of traffic on structural form are provided in reviews by Soane (1982), Koolen and Kuipers (1983), and Hakansson et al. (1988).

Roots can alter the structural form when root growth is accomplished by expanding existing pores and thereby reducing the porosity of the soil adjacent to the root channel. The entry of a root into a pore is determined by the orientation of the pore relative to the root (Dexter, 1986a,b) and by the strength of the soil matrix surrounding the root. Roots can exert pressure in both axial and radial directions, and the maximum pressures in the two directions appear to be of comparable magnitude (Misra et al., 1986). The pressures that have been observed for different species by different investigators range from 0.2 to 1.5 MPa (Misra et al., 1986, and references summarized therein). Soil deformation is mainly by cylindrical expansion of the root and must be accommodated by the loss of an equal volume of pore space from the surrounding soil. In a loosely packed bed of aggregates, the expansion of roots that have entered aggregates may fracture the aggregates, resulting in a decrease in aggregate size, provided the radial pressure exceeds the tensile strength of the aggregates and the confining pressure imposed by the medium surrounding the aggregates. Otherwise the loss of porosity must occur by compressing the soil around the root. Guidi et al. (1985) found that the porosity of aggregates associated with roots was always lower than the porosity of bulk soil aggregates and that the loss was due to a loss in the percentage of large pores: bulk soil aggregates had about 32% of total porosity formed by pores larger than 5 μm, whereas root soil aggregates had only 18%. Dexter (1987b) has analyzed data of Greacen et al. (1968) and shown that the porosity increases exponentially with the distance from the root.

The extent to which roots grow into failure zones created by freezing or rapid wetting has not been studied. However, the same factors that control the growth of a root into a pore would be expected to apply.

The loss of porosity in soil adjacent to the root, combined with the organic materials generated in the rhizosphere, will result in soil adjacent to the root having much greater strength or stability when the root dies. The remaining channel represents an important site for water flow or root development, provided the soil remains undisturbed. An analysis (Edwards et al., 1988) of macropores at different depths in a silt loam soil in Ohio, which had been farmed for 20 years in continuous no-till corn, showed that pores $>$0.4 mm in diameter accounted for approximately 1.4% of the total area, while 80% of the pores were $<$1 mm in diameter and were assumed to be root channels. The increased number of pores that was observed at lower depths implied greater persistence of the root channels.

The mechanical effects of roots on the structural form are complemented by the removal of water from the soil by transpiration and the consequent effects arising from an increase in the amplitude of wetting–drying cycles. It is therefore not surprising that the effect of root growth on aggregation depends, as for wetting–drying and freezing–thawing, on the initial state of the soil. Plant roots have been observed to cause a decrease in aggregate size or an increase in aggregate size, depending on whether the soil was originally compact or loose, respectively (Bui Huu Tri, 1968).

The susceptibility of soil to changes in structural form by the compressive and shearing forces exerted by hooves of livestock would be expected to be similar to that due to other forms of traffic. Warren et al. (1986) found that high stocking rates of cattle (2.7 ha per animal unit per year) reduced infiltration rates and increased soil bulk density at a depth of 0 to 50 mm in a silty clay soil in Texas. Reductions in infiltration were smaller at lower stocking rates and when trampling occurred when the soil was drier. A more complete review of the influence of livestock on the hydrologic properties of soil is provided in Warren (1987).

4. Soil Faunal Activity

Soil fauna, such as earthworms, can alter the structural form by the creation of burrows (Kladivko et al., 1986; Zachman et al., 1987). The nature of the burrows formed will depend on the species. Species that feed on surface litter create burrows that are more or less permanent refuges. These burrows are usually vertical for most of their length, sometimes branch to several entrances, and are lined with mucous (Lee, 1985). Geophagous species feed on organic material in the soil and have more extensive burrows that are predominantly horizontal, with some vertical components and openings to the surface. Their burrows are often filled with casts (Lee, 1985). Earthworms form burrows by eating their way through soil and pushing through crevices (Edwards and Lofty, 1972). McKenzie and Dexter (1988) have measured maximum radial pressures of about 0.2 MPa that were exerted during burrowing by the species *Aporrectodea rosea*. Other studies did not reveal compaction around earthworm burrows, leading McKenzie and Dexter to conclude that mechanisms that the earthworm employs to overcome confining stresses which may limit its forward movement include the use of radial pressure to weaken failure zones ahead of itself.

The rate of burrow formation by earthworms would be expected to be controlled by the population dynamics. The ecology of earthworms is influenced by soil factors such as pH, moisture, temperature, aeration, soil type, and food supply (Edwards and Lofty, 1972) as well as by soil management practices such as tillage, mulching, fertilization, and the use of biocides (Lee, 1985). It is therefore not surprising that seasonal variations in burrow systems exist (Kretschmer, 1982). However, there is little information on the relation between population dynamics and the formation or the persistence of burrows (Gibbs and Reid, 1988).

B. Structural Stability

The resistance of the structural form of soil to change when different stresses are applied can be characterized for different forms of stress and at different scales of structural form. Regardless of what stability parameter is being considered, however, the structural form at smaller scales (such as microaggregates and micropores) tends to be most stable (Tisdall and Oades, 1982; Gregorich et al., 1988), and the stability declines with increasing scale (Coughlan et al., 1973;

Katou et al., 1987). Agricultural management practices have their greatest impact on the structural stability at the scale of macroaggregates (Tisdall and Oades, 1980b) and macropores [although, as noted by Gibbs and Reid (1988), there is very little information on the stability of macropores].

1. Role of Failure Zones

Deformation of the structural form under an applied stress will occur at the weakest points in the matrix, and application of increasing stress results in failure in zones of increasing strength. A discussion of the processes leading to changes in stability can be most effectively approached by considering processes that lead to a change in the distribution of strengths of the failure zones with emphasis on the weakest failure zones. The distribution of strengths in failure zones in soil can be illustrated by considering the distribution of strengths of elemental volumes. The term *elemental volumes* is used in this context as a mathematical concept and has no structural significance, that is, there are no discontinuities in strength between elemental volumes. The strength within an elemental volume will be considered, for simplicity, to be constant and to reflect the ability of the solid material in that elemental volume to stick together as a consequence of one or more mechanisms such as effective stress, enmeshment by strands of stabilizing materials, electrostatic attraction, or interatomic bonding. The strength of each elemental volume will be equated to the energy required to create a surface extending across the volume. A group of elemental volumes will have a statistical distribution of strengths. The actual distribution of strengths of the elemental volumes is not known. Two possible distributions are given in Figure 5 and are

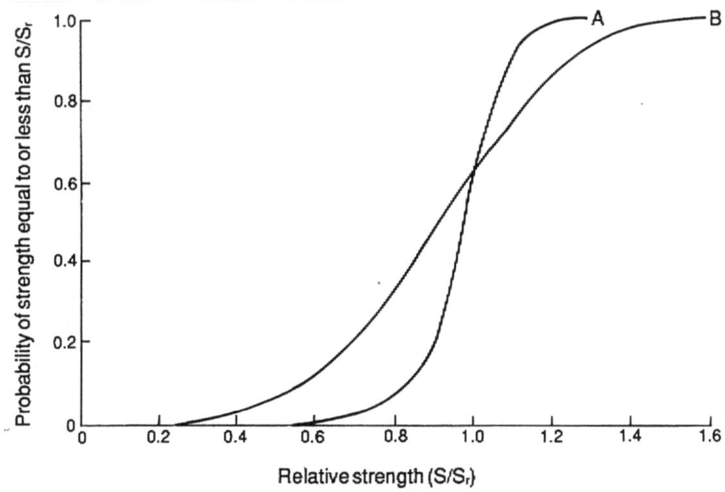

Figure 5. Diagrammatic representation of the distribution of relative strengths of elemental volumes of soil.

presented only for illustrative purposes. The strength of the elemental volumes is defined relative to a reference strength that is characteristic of the soil in order to provide reduced values. The reference strength is the strength at which the probability variable on the vertical axis equals 0.632. The distributions differ in the dispersion of strengths of the elemental volumes with distribution *B* having the wider distribution of strengths.

A failure zone implies an association or grouping of elemental volumes of low strength, and the strength of the zone would reflect the distribution of strengths in the elemental volumes, the connectivity or distance between zones of low strength, and the strengths of elemental volumes in this intervening distance. The strength of the failure zone is the energy required to create a new unit of surface area or propagate a crack (Skidmore and Powers, 1982) and has been called the *apparent surface energy* (Hadas, 1987a). A soil clod (or aggregate, microaggregate, etc.) would be expected to contain failure zones with a range of strengths. The stability of the soil structure then relates to the distribution of the strengths of the weakest failure zones. A structure will be completely stable if the stress that is applied is less than the strength of the failure zones. The probability of encountering a zone of low strength increases with the distance traversed through the soil, and therefore the larger the scale being considered, the lower the stability. The existence of failure zones is implicitly recognized in descriptions of brittle fracture, and this theory has been used by Braunack et al. (1979) and Hadas (1987b) in an analysis of the tensile strengths of aggregates. The tensile strength is the minimum stress at which the stress-induced failure in one or more of the elemental volumes in the network extends to form a continuous failure surface. Utomo and Dexter (1981c) have utilized measurements of the variation in tensile strength of aggregates to define the soil characteristic friability; the greater the dispersion in the strengths of the failure zones, the greater the friability and the greater the range in aggregate sizes produced on tillage. The type of distributions illustrated in Figure 5 highlight distributions of extremes, largest or smallest, of large samples and are known as the third asymptotic distribution of smallest values (Gumbel, 1960). This distribution has been used by Braunack et al. (1979) to describe the distribution of tensile strengths of clods or aggregates during brittle fracture. The dispersion or spread of these distributions is defined as the friability by Utomo and Dexter (1981c). Curves *A* and *B* in Figure 5 represent friabilities of 0.10 and 0.25, respectively. On the basis of practical experience, Utomo and Dexter (1981c) proposed that soils with friabilities less than 0.05 are not friable, and values of 0.05 to 0.10, 0.10 to 0.25, 0.25 to 0.40, and greater than 0.40 reflect slightly friable, friable, very friable, and unstable conditions, respectively. The influence of cropping practices on friabilities measured in this way is illustrated in values of 0.08 and 0.26 for cultivated and permanent pasture treatments on a hardsetting soil in Australia (Chan, 1989a). Although the characterization of failure zones is most advanced in theories related to tensile strength, the concept is particularly valuable when considering the stability of the structural form of soil to other kinds of stress and the impact of agricultural practices on structural stability.

Processes that alter the strengths of failure zones can be considered from the perspective of the influence of the processes on the strength of elemental volumes. The strength of elemental volumes may be changed by changing the volume fraction of solids in these volumes or by altering the strength or number of bonds between the surfaces of particles. The spatial relation of elemental volumes that have had a change in strength may also influence the connectivity between volumes of low strength. The strength of the failure zone will change with water content and can be modified by drying, wetting, freezing, thawing, tillage, traffic, and biological activity.

An increase in the volume fraction of solids in elemental volumes can occur by compaction through tillage or traffic or by the illuviation of clay, and will lead to the strengthening of failure zones. Loss of porosity has been associated with an increase in the stability characteristics such as tensile strength (Dexter et al., 1984; Angers et al., 1987) and wet aggregate stability (Mokhtaruddin and de Boodt, 1983). An increase in the volume fraction of solids in an elemental volume will also be associated with shrinkage by drying, freezing, and compression around roots. However, in these cases the increase in strength of some elemental volumes must be accompanied by a loss of strength in other elemental volumes where the volume fraction of solids decreases (i.e., where root channels and cracks are located). The strength and spatial distribution of the failure zones will, therefore, depend on the connectivity of zones of low strength. For instance, freezing and the formation of ice lenses, as illustrated in Figure 4, will produce failure zones that are continuous at a macroscopic level. It is not surprising then that freezing results in a decrease in tensile strength (Utomo and Dexter, 1981c), resistance to penetration (Voorehees, 1983), shear strength (Douglas, 1986), and the stability under wet sieving of aggregates that were larger than 250 μm (Willis, 1955; Sillanpaa and Webber, 1961; Bullock et al., 1988).

The strength of failure zones is strongly influenced by water content. Decreasing water potential results in the progressive drainage of large pores, increased curvature in the air–water menisci of smaller pores which are water-filled, and increased effective stress. Therefore elemental volumes containing drained pores will be weaker than volumes containing pores filled with water and under tension. The influence of changing water content on the strength of failure zones will depend on the variation in pore size. The water content may also influence the strength of bonds between mineral particles, as will be discussed later in this section. Responses to changing water potential have been observed for aggregate stability (Francis and Cruse, 1983), tensile strength (Guerif, 1988; Grant, 1989), and resistance to penetration (Mirreh and Ketcheson, 1973). The water content varies continuously under field conditions, and therefore structural stability changes continuously. The relation between stability and water content would be expected to change with cropping systems whenever the changes in cropping systems cause changes in pore size distribution, changes in the amount or spatial distribution of water extracted, or changes in the sensitivity of cementing or stabilizing materials to changes in water content.

The spatial distribution of zones of weakness will determine the size and shape of soil aggregates when a given stress is applied to a soil clod, and the distribution of strengths of elemental volumes in a failure zone will determine the roughness of surfaces created on failure of the zone. An analysis of the shape of aggregates from the arable layer of different Dutch and Australian soils suggests that aggregates are less round with increasing clay content and decreasing organic matter content (Dexter, 1985). The roughness of fracture surfaces diminishes as soils dry (air-filled pores which represent low strength areas are closer together), and in the case of soils with water contents that are similar and below the lower plastic limit, the roughness of fracture surfaces decreases with decreasing organic carbon content (Grant, 1989).

Changes in agricultural practices will be manifested first in macropores and other zones of lowest strength. It is here that solute transport occurs most readily, roots develop, deformation will occur most readily during tillage, crop residues and soil amendments that are incorporated into the soil will be located, and where microbial activity is concentrated. It is not, therefore, surprising to find that such zones differ from the bulk soil not only in strength characteristics but in chemical characteristics as well (Hantschel et al., 1988).

2. Stabilizing Materials in Failure Zones

Materials responsible for bonding soil particles together may be inorganic or organic in nature. Detailed reviews on the nature of stabilizing materials and the scale at which they operate have been provided by Emerson (1983), Lynch (1984), Oades (1984), and Burns and Davies (1986), and therefore comments on stabilizing materials in this chapter will be brief and restricted to those stabilizing materials which change in quantity or effectiveness in response to agricultural practices.

The location of stabilizing materials in failure zones will determine their effectiveness and, in the case of organic stabilizing materials, their persistence. The role of the location in the persistence of organic stabilizing materials will be considered later in this section. The importance of the location relative to effectiveness depends on the dimensions of the stabilizing material relative to the distance between mineral particles. Inorganic stabilizing materials and organic molecules will be capable of bonding clay particles and cementing larger mineral grains together at their points of contact. Such contact points may exist in the soil matrix and at contact points between adjacent faces of microcracks. However, where grains are separated by greater distances, such as might be encountered in noncontact zones of microcracks or cracks, materials of a molecular scale will be largely ineffective and binding must occur by materials such as roots and fungal hyphae.

The development of a strongly bonded mineral-organic matrix in wet soils containing clay is contingent on the clays remaining in a flocculated state; as clays become more dispersed, the stability of wet soils decreases. The importance of the interaction between clay particles on the strength of the mineral-organic matrix is illustrated in Emerson's diagrammatic representation of possible arrangements of clay domains, organic matter, and quartz in a soil crumb (Emerson, 1959). Shanmuganathan and Oades (1982) found that the wet aggregate stability of particles 50 to 250 μm in diameter decreased six- to sevenfold when the dispersible clay content increased about tenfold. However, as drying proceeds, the dispersed clay, which is relatively mobile, can reorient or accumulate at points of contact between larger particles, and in the absence of water the clay particles can function as cementing materials. Consequently soils with higher dispersible clay content in wet conditions can have higher strength characteristics such as modulus of rupture, tensile strength, and lower friability when dry (Shanmuganathan and Oades, 1982; Chan, 1989b). In contrast, flocculation of clay leads to greater strength when wet and lower strength when dry (Ingles, 1962).

The degree of flocculation or dispersion can be changed by altering the surface charge of the clays (Schanmuganathan and Oades, 1982); by altering the composition of the pore fluid or the composition of exchangeable cations (Rengasamy et al., 1984; Rengasamy et al., 1986), or by mechanical forces such as tillage or raindrop impact (Emerson, 1983). A relation between sodium absorption ratio, concentration of cations in solution, and dispersibility of clays in Australian red-brown earths has been proposed by Rengasamy et al. (1984) and is illustrated in Figure 6. Clay in dispersive soils will disperse spontaneously on wetting, whereas soils that are potentially dispersible require some degree of mechanical manipulation before the clay becomes dispersed. Flocculated soils are unlikely to release clay even when exposed to extensive shearing or molding. Although the relations in Figure 6 may change significantly in soils of different mineralogy, organic matter content, or the presence of inorganic cementing materials (Rengasamy et al., 1984; Ali et al., 1987), they illustrate how the dispersibility of clay may change in zones of weakness in response to changes in

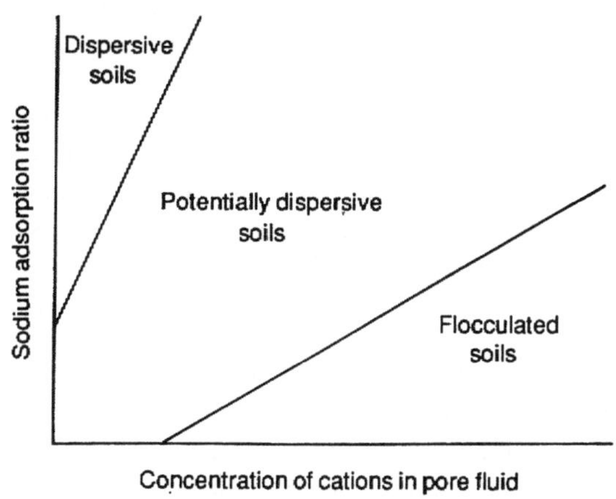

Figure 6. Diagrammatic representation of the relation between factors influencing the dispersibility of clay. [Adapted from Rengasamy et al. (1984).]

electrolyte concentrations in the pore fluid, composition of exchangeable ions, and mechanical disruption. Agricultural practices that may bring about changes in the dispersibility of clay include the use of liming materials, fertilizers, changes in the quality of irrigation water, and intensity of tillage.

Amorphous oxides of iron or aluminum and precipitated aluminosilicates can function as cementing materials, thereby reducing the dispersibility of clay (see review by Emerson, 1983) and contributing to bonds between other mineral particles directly or through involvement with organic molecules. The solubility of these materials is pH-dependent, and therefore the stability of soils can be changed by changing the soil pH (Yeoh and Oades, 1981). However, there is little information on the influence of management practices on the effectiveness of amorphous inorganic materials as stabilizing agents.

Organic materials play a major role in the structural stability of topsoil. The rate and direction of change in organic matter content varies with agricultural practice, initial organic matter content, and climate (Johnston, 1986; Mann, 1986). Decreasing organic matter contents have been correlated with a loss of stability when a range of stability characteristics are considered: wet aggregate stability (see summary by Tisdall and Oades, 1982), dispersibility of clay (Emerson, 1954), tensile strength (Guerif, 1988; Grant, 1989), and shear strength (Davies, 1985). Organic matter may contribute to stability directly as a cementing material or indirectly through its effect on porosity or water content (Grant, 1989). Organic stabilizing materials can complement the effectiveness of inorganic stabilizing materials. Greenland et al. (1975) noted that English and Welsh soils were most likely to be unstable when the organic carbon content was less than 2% and there was no free $CaCO_3$ present. Muneer and Oades (1989a, 1989b, 1989c) showed that the combined addition of calcium in the form of gypsum or agricultural lime with an organic substrate (glucose or wheat straw) was able to reduce dispersion and increase the proportion of aggregates greater than 250 μm to a greater extent than the addition of any of these materials by themselves. The increased stability was also observed to persist for a longer period of time and seemed to be associated with the formation of Ca-organic complexes.

Soil organic matter is a continuum of materials that can be represented by pools which decompose at different rates (Jenkinson and Rayner, 1977). The pools include decomposable plant material, resistant plant material, soil biomass, physically stabilized organic matter, and chemically stabilized organic matter fractions. The chemical composition or quality of carbon and the degree of physical protection of the carbon against biological degradation are considered by many authors to distinguish between the carbons in the different pools (see Bosatta and Agren, 1985; Parton et al., 1987). Major changes in stability can occur before significant changes occur in the amount of organic matter in soil (Baldock and Kay, 1987), suggesting that structural stability may be more strongly influenced by the carbon in some pools than that in others and that the level of carbon in these pools responds more rapidly to changes in agricultural practice than the total carbon pool.

Organic materials which contribute to bonding between mineral particles range from short-chain organic acids to plant roots and therefore vary in dimensions over four orders of magnitude (Figure 1). The size of the stabilizing material will obviously influence the connectivity between elemental volumes. Dispersion of clay particles is promoted by the adsorption of anions of organic acids, which increases the negative charge on clays (Shanmuganathan and Oades, 1983a). The acids are produced by plants, bacteria, and fungi. Dispersion of clay has also been attributed to materials in root exudates capable of complexing iron and aluminum, which form bridges between organic matter and the mineral surfaces (Reid et al., 1982). The tendency to dispersion is offset by the binding action of polysaccharides, mainly mucilages produced by bacteria, but also by plant roots and fungal hyphae (Oades, 1984). The strength of bonds is enhanced by multivalent cations, which act as bridges between organic colloids and clays (Oades, 1984). Roots and fungal hyphae enmesh soil around failure zones, thereby increasing the strength of the failure zones.

Roots will be located in the weakest failure zones and in rigid pores that are larger than about two-thirds of the diameter of the root when grown under zero stress conditions (Dexter, 1988). Root hairs and fungal hyphae are, respectively, one and two orders smaller in diameter than roots (Figure 1) and will therefore penetrate much smaller pores in the adjacent soil. Tisdall and Oades (1980b) report hyphal lengths of up to 19 m/g soil compared to root lengths of 1 m/g soil under different crops. The root hairs and fungal hyphae therefore provide an effective mechanism of dispersing carbon well beyond the position of the root and the immediate rhizosphere [mucigel of plant and microbial origin has been observed 50 μm beyond root epidermal cells (Emerson et al., 1986)]. Enmeshment and the deposition of carbon from root exudates, fungal hyphae, and root hairs in pores smaller than those in which the roots are located will increase the strength of weaker failure zones and contribute to the stabilization of macroaggregates (Oades, 1984). The stabilizing materials in such pores from which plants can remove water (see Figure 1) are of particular importance in increasing the stability of those soils that contain a large proportion of pores in this size range. Quirk and Williams (1974) showed that the mechanical strength of aggregates from a red-brown earth was more than 50% higher when polyvinyl alcohol penetrated pores with an effective radius of 15 to 50 μm than when it was restricted to pores with radii of 7.5 to 15 μm or 50 to 150 μm.

The decomposition of stabilizing materials will be occurring simultaneously with the formation of new stabilizing materials. The amount of a given material present at any time will reflect the rate of addition relative to the rate of loss. Tisdall and Oades (1982) have suggested that organic stabilizing materials fall into three broad classes based on their persistence or rate of loss: (1) transient (mainly polysaccharides), (2) temporary (roots, fungal hyphae and bacterial cells), and (3) persistent. The persistence of transient materials is on the order of weeks, whereas temporary materials may exist for periods varying from months to a few years (Tisdall and Oades, 1982). Persistent materials remain in the soil for much longer periods and may account for 52 to 98% of the total organic matter in soils (Tisdall and Oades, 1982). The relative stability of persistent organic matter in some soils may be due to physical association with the inorganic components of the soil rather than an inherent chemical or biochemical inertness of the organic matter itself (Tiessen et al., 1984; Skjemstad et al., 1986; Oades et al., 1988), although in other soils distinct chemical differences have been reported (Skjemstad and Dalal, 1987).

Transient, temporary, and persistent stabilizing materials can coexist in failure zones. The growing roots will be releasing extracellular carbon into the rhizosphere, and this may account for one-quarter to one-third of the photosynthate deposited below ground (Barber and Martin, 1976; Helal and Sauerbeck, 1984). This material is substrate for a large microbial population. Soil microbial biomass increases with the root growth and rooting density of the crop (Lynch and Panting, 1980). Following rhizodermal differentiation, exudate release ceases, microbial cells lyse, and the soluble lytic products and the excretory products of browsing protozoans return to the root in the transpiration stream (Foster, 1988). Much of the mucilages produced by the root and microorganisms are mineralized and hence are considered temporary binding agents. However, some of this organic material along with fragments of roots and fungal hyphae may interact with clay and become physically protected against further decomposition (Jenkinson and Rayner, 1977; Elliott et al., 1980) and may therefore become the center of microaggregates (Oades, 1984). Bacteria which have entered pores smaller than 5 μm in diameter may escape predation by soil protozoans and the effects of fungal antibiotics and may become completely enclosed in a clay fabric due to clay swelling or compaction through growth of nearby roots (Foster, 1988). Polysaccharides that diffuse or get squeezed into pores smaller than 1 μm will be protected since they will not be accessible to attack by soil organisms (Adu and Oades, 1978). Hetier et al. (1986) showed that about half of the nonroot carbon that is not mineralized is associated with mineral particles <5 μm in diameter and is not soluble at pH = 10 or in 1*M* HCl. A higher proportion of root-originating carbon is fixed in clay-rich soil, and a decreasing proportion is incorporated into soil microorganisms (Merckx et al., 1985). These materials would then be considered persistent stabilizing materials, and the strength of elemental volumes made up of these materials would be expected to increase.

Organic stabilizing materials that are physically protected against biological degradation can become accessible and mineralized as failure zones are created by tillage and by wetting and drying (Rovira and Greacen, 1957; Powlson, 1980). Fracturing of high-strength zones would be expected to occur when they are isolated between zones of low strength. Extensive tillage and a diminished rate of replenishment of transient and temporary stabilizing materials result in a progressive loss of these organic stabilizing materials, weakening of failure zones, and a subsequent decrease in aggregate size.

Soil fauna, such as earthworms, play an important role in the comminution of organic materials and the redistribution of organic matter in soil. In addition, the passage of organic materials and soil particles through the gut of the earthworms results in an intimate association between organic materials and mineral materials and changes the strength of bonds between mineral grains. The stability of earthworm casts depends to a considerable extent on the concentration and type of organic matter ingested (Shipitalo and Protz, 1988), the microbial activity in the casts after they are produced (Lee, 1985), and the time available for thixotropic hardening and bonding of clay with incorporated organic debris (Shipitalo and Protz, 1988).

A model for the organization of aggregates in soils in which organic matter is the main binding agent has been proposed for red-brown earths in Australia (Tisdall and Oades, 1982) and may apply more generally to grasslands soils (Elliott, 1986). Aggregates greater than 2000 μm in diameter are proposed to be held together by a network of roots and fungal hyphae and are made up of microaggregates 20 to 250 μm in diameter. Microaggregates 20 to 250 μm in diameter are considered stable to rapid wetting, are not destroyed by agricultural practices, and consist largely of particles 2 to 20 μm in diameter bonded together by persistent organic materials and crystalline oxides. The distinction between microaggregates and macroaggregates is based on the susceptibility to slaking, and the upper limit to microaggregate size, which is defined in this way, defines the minimum separation of failure zones created by slaking. The size limit would be expected to vary depending on the spatial relations of failure zones produced by shrinkage, swelling, and ice crystals or ice lenses of different morphology and was noted by Tisdall and Oakes (1982) to be much different for red-brown earth and black earth soils in Australia.

The hierarchical nature of the soil structure model proposed by Tisdall and Oades (1982) and its relation to failure zones of different strengths can be inferred from the variation with distance in the resistance to penetration experienced by a needle penetrometer (150-μm diameter) as it traverses aggregates (Grant et al., 1985). Minima in the profiles in tip resistance (Figure 7) indicate zones of low resistance within macroaggregates and are compatible with the concept of the macroaggregate being a grouping of microaggregates separated by zones of different strength. Groenevelt et al. (1984) have analyzed this type of data obtained on intact cores by calculating the percentage of linear path

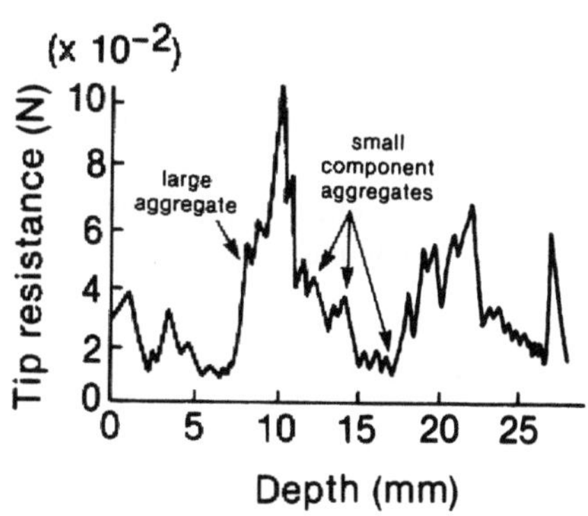

Figure 7. Profile of resistance to penetration experienced by the tip of a needle penetrometer penetrating beds of aggregates varying from 6.4 to 12.8 mm in diameter. [From Grant et al. (1985).]

traversed in which the tip pressure exceeds an arbitrary limit; the remainder represents the cumulative proportion of the linear path made up of zones of strength lower than the arbitrary limit. A hierarchical nature of the soil structure is also illustrated in tensile strengths of air-dry aggregate from soils with different cropping histories (Hadas, 1987b).

3. Seasonal Variations

Structural stability would be expected to vary seasonally as physical processes related to climate, plant growth, tillage, and traffic create new failure planes or weaken existing ones. Compression may provide compensatory strengthening of failure zones. Plant growth will alter the supply and distribution of soil carbon in failure zones, and soil organisms will process new and existing organic stabilizing materials. Microbial activity is strongly influenced by variations in soil water content during the growing season (McGill et al., 1986; Doran et al., 1988) and soil temperature (van der Linden et al., 1987). Desiccation can lead to increased orientation of clay-sized mineral particles or accumulation of clay-sized particles in the menisci between larger particles (Brewer and Haldane, 1957; Horn and Dexter, 1989), a change in the stabilizing behavior of organic molecules (Reid and Goss, 1982), and the crystallization of solutes when the concentration of the solutes in the pore fluid is close to the solubility limit prior to the onset of drying (Gifford and Thran, 1974; Kemper et al., 1988). Changes caused by drying may not be completely reversible, or only slowly reversible on rewetting, and may account for the dependence of wet aggregate stability on the moisture content when slaking is minimized (Churchman and Tate, 1987; Gerard, 1987; Perfect et al., 1989a). Structural stabilities measured at different times of the year reflect an integration of all of these processes. Seasonal variation has been observed in widely different environments in wet aggregate stability (Alderfer, 1946; Stefanson, 1968; Ramsey et al., 1986; Coote et al., 1988; Bullock et al., 1988), dry aggregate stability (Bisal and Ferguson, 1968), and shear strength (Douglas, 1986; Coote et al., 1988). The seasonal fluctuations in stability are often greater than changes due to management and can obscure changes caused by management (Perfect et al., 1989a).

The dynamic nature of failure zones when biological activity, compaction, and freezing/thawing occur singly or in combination has been illustrated under laboratory conditions using colored glass beads to label artificial soil aggregates (Terpstra, 1989). Breakdown and reconstitution of these aggregates appeared to occur relatively quickly under laboratory conditions and provided convincing qualitative evidence of the strengthening and weakening of failure zones which must occur continuously under field conditions.

C. Structural Resiliency

The ability of a soil to recover its structural form through natural processes when external stresses are reduced or removed is particularly important when the external stresses are resulting in a structural form that is unsuitable for plant growth and exacerbates the detrimental impact of agricultural practices on the environment. Field experience suggests that soils differ considerably in their resiliency. Such differences relate to both the extent that the structural form is altered by wetting, drying, freezing, and the activity of soil fauna, and the persistence of these changes (i.e., the structural stability).

Soils that are most resilient are those known as self-mulching soils. The structural form of such soils is not particularly sensitive to mechanical processes. If much of the macroporosity is detroyed when the soil is tilled when wet, a desirable structural form can be recreated during wetting and drying or freezing and thawing cycles. Blackmore (1981) has characterized self-mulching soils as those which are soft when wet, do not disperse, but have a marked tendency to shrinking and swelling and have a crumb density >2 g cm^{-3} when dry. Such soils might be more appropriately called self-tilling soils (Blackmore, 1981).

Other soils, while less resilient than self-mulching soils, can still exhibit an improvement in structural form through the creation of microcracks (Utomo and Dexter, 1981a) and the formation of biopores (Shipitalo and Protz, 1987). The resiliency of soil would be expected to decrease rapidly with depth in the soil profile as a consequence of a decrease in the rate of wetting and the extent of both drying and freezing. Bullock et al. (1985) found that the porosity in the top 5 cm of soil compacted by wheel traffic was regenerated within 18 months with most of the improvement occurring during the summer months. However, the regenerated structure mainly consisted of planar voids, unlike the unwheeled soil, which was characterized by packing voids and biopores. Soil at greater depths remained compacted for longer. Blake et al. (1976) recorded no regeneration of the porosity of a clay loam soil in Minnesota at 30 to 49 cm nine years after subsoil compaction had occurred, although the soil would annually freeze to greater depths. Similar observations in the Netherlands (van Ouwerkerk, 1968) and in Sweden (Hakansson, 1985) may reflect the failure of pores created during freezing to persist in some soils after thawing (Kay et al., 1985).

The persistence of the structural form is related to the structural stability of soil, and, as noted above, external forces can alter stability. Therefore, recovery of stability when the external forces are removed or reduced can contribute to the recovery of the structural form. Molding soil introduces a large number of failure zones and results in a dramatic loss in stability. Subsequent to molding soils, increases in several stability characteristics have been reported: shear strength (Davies, 1985), dispersibility of clay (Molope et al., 1985), wet aggregate stability (Kemper and Rosenau, 1984), modulus of rupture (Kemper and Rosenau, 1984), resistance to penetration (Utomo and Dexter, 1981b; Molope, 1987), tensile strength (Utomo and Dexter, 1981b), and compression resistance (Utomo and Dexter, 1981b). Mitchell (1960) suggested that during molding part of the shearing energy is utilized to disperse the clay particles. When shearing ceases, the structure adjusts into a lower free energy. Utomo and Dexter (1981b) have noted that such behavior would require an interparticle balance of forces such that soil will flocculate if given a chance, and the flocculation tendency is not so strong that it cannot be overcome by mechanical action, such as shearing or stirring the soil. Following this argument the loss of strength and subsequent recovery would be greatest when clay in the failure zones is potentially dispersive (Figure 6).

The extent of strength recovery can vary. Mitchell (1960) defined thixotropy as an isothermal, reversible, time-dependent process occurring under conditions of constant composition and volume, whereby a material stiffens while at rest and softens or liquifies when molded. Utomo and Dexter (1981b) have employed the term *age hardening* to describe the increases in strength that do not meet Mitchell's criterion.

The increase in strength depends on moisture content, biological activity, and organic matter content. Strength increases most rapidly at water contents close to the plastic limit (Utomo and Dexter, 1981b). However, the reasons for this relationship are not clear. It may be speculated that in drier soils the higher concentration of cations may maintain the clay in failure zones in a flocculated state, and therefore thixotropic regain would be negligible, whereas at high water contents low electrolyte concentration would cause the clay to disperse on its own accord and thixotropic regain would also be negligible. The range in water contents considered by Utomo and Dexter (1981b) would result in concentration changes by no more than a factor of 2 and would therefore require that the concentration of the pore fluid be close to the critical coagulation concentration for flocculation–dispersion to be so sensitive to solute concentration. Under saturated conditions Kemper et al. (1987) observed a slow decline in stability with time (which was attributed to the solubilization of slightly soluble compounds that were helping cement aggregates) followed by an increase in stability (which was attributed to component ions, atoms, and molecules migrating to lower energy positions where they are more effective in bonding particles together). Biological factors such as prolific growth of fungi and an increase in bacterial populations have been observed after molding and coincide with initial increases in stability (Molope and Page, 1986), and the biological factors can supplement stabilization by inorganic materials (Utomo and Dexter, 1981b). An analysis of the change in resistance to penetration over time under different compaction pressures suggests that the increase in strength of some soils is primarily due to an increase in the number of particle–particle bonds, whereas in other soils the increase in strength appears to be due primarily to a reinforcement of particle–particle bonds by a cementation mechanism (Dexter et al., 1988). Age hardening by the second mechanism may be inhibited by soil organic matter (Dexter et al., 1988).

There are several practical implications of thixotropic or age hardening behavior. Arya and Blake (1972) noted an increase in wet aggregate stability after plowing, with most of the increase occurring in the first 10 to 12 hours. Soils would therefore be expected to be less susceptible to slumping and erosion the greater the length of time between tillage and a rainfall event. However, the increase in stability with time is unlikely to compensate for the loss in stability associated with the loss of organic stabilizing materials, which may be associated with some cropping practices (Molope et al., 1985).

IV. Managing Soil Structure with Cropping Systems

Soil structural form, stability, and resiliency can be managed through cropping practices (species selection and crop management), tillage and traffic control, drainage, and soil amendments. The impact of management practices on soil structural characteristics will vary with the soil and climate, and for any given soil and climate combination they will vary with depth in the profile.

A. Cropping Practices

The characteristics of the plant species being grown, sequence of different species, and frequency of harvesting are aspects of cropping practices that influence the formation of biopores (by roots and soil fauna) and the amount and distribution in the soil of organic stabilizing materials.

The creation of biopores by roots is most important in soils that are compacted at depths not readily accessible to tillage equipment. For instance, Elkins et al. (1977) have shown that bahiagrass (*Paspalum notatum* Flugge "Pensacola") roots will penetrate soil layers that mechanically impede cotton *(Gossypium hirsutum)* roots. Bahiagrass increased the number of pores greater than 1.0 mm in diameter, and this facilitated increased rooting densities of subsequent cotton crops to a depth of at least 60 cm. The biopores were sufficiently stable that cotton grown where bahiagrass sod had been plowed less than three years before still yielded more than cotton grown where the compacted zones were loosened by chiseling. The effectiveness of different species in penetrating compacted soil would be expected to be related to the root morphology and to the maximum radial and axial pressures that can be exerted, but this has not been investigated in detail.

Crops provide the food supply for biopore-forming soil fauna such as earthworms. The choice of crops and associated tillage can, therefore, influence the quantity and quality of food available and consequently population dynamics. For instance, Low (1972) reports populations of 1.25 to 1.8×10^6, 1 to 2×10^5, and 4 to 7.5×10^5 worms/ha for old grassland, old arable land, and grassland converted to arable land for 3 years, respectively. Shipitalo et al. (1988) have speculated, on the basis of laboratory studies of the diet and casting activity of surface-feeding species, that food supply may limit the populations of these species in alfalfa *(Medicago sativa)* fields, whereas food quality may limit population levels in fields used for corn *(Zea mays)* production.

The effectiveness of different crops in improving the structural form and stability at different depths is related to the variation with depth in the amount of water extracted, the total amount of photosynthate deposited by the plant, and the persistence of this carbon in the soil. Rooting densities tend to decrease exponentially with depth in well-structured soils with the extent of decrease being specific for each crop, growth stage, and soil (Hamblin, 1985; Dwyer et al., 1988). Most stresses on soil structure are imposed at or near the ground surface, and therefore the influence of plants on the soil structure is very important in this zone.

The total amount of photosynthate deposited below ground and the resistance of this material to complete mineralization vary considerably between species. For instance, the total amount of rhizodeposition by bromegrass *(Bromus inermis)* can be twice as large as that for corn (Davenport and Thomas, 1988), and carbon originating from bromegrass shows greater persistence than that from corn (Davenport et al., 1988).

The root length density in the top 15 cm of soil generally increases in the following order for different types of crops: row crops [corn, soyabean *(Glycine max)]* < cereals (barley, winter wheat) < grasses [bromegrass, blue grass *(Poa pratensis),* western wheatgrass *(Agropyron smithii)]*. Studies in which the influence of the crop species on the stability of surface soils is not confounded with tillage or initial organic matter content indicate the same sequence for increases in root length as for increases in wet aggregate stability (Duffey et al., 1986; Stone and Buttery, 1989; Perfect et al., 1989b) and dispersible clay (Perfect et al., 1989b).

Considerable intervarietal differences exist in root characteristics (Scott Russell, 1977), suggesting that breeding programs could enhance the effectiveness of varieties to improve the soil structure. This would require the introduction of criteria for varietal selection which are not currently involved in breeding programs.

The extent of structural stabilization by a crop may depend on the management of the crop during the growing season. For instance, Tisdall and Oades (1980a) observed a more rapid increase in wet aggregate stability when ryegrass *Lolium perenne),* grown under glasshouse conditions, was clipped at monthly intervals. Clipping did not increase root length, but the authors noted that cutting appeared to stimulate the growth of mycorrhyzal hyphae, although the effect was not statistically significant. Perfect et al. (1989b), however, were unable to see any difference in structural stabilization by red clover *(Trifolium pratense)* cut once or twice during the growing season. Although a great deal of emphasis has been placed on the effectiveness of different crops in improving the soil structure, the persistence of these effects has been documented in only a small number of studies (Low, 1972; van Bavel and Schaller, 1950).

Plant species that regenerate the structural form and stability can be incorporated into cropping systems by being interseeded in the commercial crop and retained or by being planted as a cover crop between regular cropping seasons (Bennett et al., 1976; Scott et al., 1987). Species that regenerate the soil structure may also be used as alternatives to weed fallow or bush fallow utilized under shifting cultivation in some tropical soils (Lal et al., 1979). Particular emphasis is now being placed on legumes which provide nitrogen to the succeeding crop as well as conferring structural benefits (Power, 1987). It is perhaps useful to note parenthetically that structure-regenerating crops are also often introduced during periods of depression in agriculture. Johnston (1986) has noted that typical farm soils in the United Kingdom have been in continuous grassland during periods of agricultural depression which have lasted 20 to 30 years and which have recurred during the last 200 years.

B. Tillage and Traffic

The impact of tillage on the structural characteristics can be managed through the type of tillage equipment, the soil conditions under which it is used, and the extent of use. The influence of different types of tillage equipment on soil structural characteristics has been reviewed in Bowen (1981) and will not be considered in this report.

The soil water content is the most important parameter in determining the influence of tillage on soil structure. Soils are most friable when tillage is carried out at water contents at or slightly below the lower plastic limit (Ojeniyi and Dexter, 1979a). At higher water contents soil is readily deformed but does not crumble. The ratio of plastic limit (PL) to field capacity (FC) provides a useful index of the workability of soil. Soils with a value of PL/FC approaching unity will reach optimum conditions for fracturing as soon as water draining under the influence of gravity is lost, whereas soils with a much smaller ratio will not reach optimum conditions until further drying occurs. The ratio of PL/FC can be increased by increasing the organic matter content, the critical level of which depends on the lime status of the soil (Boekel, 1965, as cited in Dexter, 1988).

Tillage causes sorting of aggregates, with smaller aggregates tending to sink to the bottom of the tilled layer and the larger ones tending to rise to the surface. The distribution of pore sizes with depth is modified as a consequence. The variation in structure with depth is greater when tillage occurs at water contents above the plastic limit than where the water content is less than the plastic limit (Ojeniyi and Dexter, 1979b).

The soil water content influences the dispersibility of clay when tillage occurs. The amount of clay dispersed increases with the water content once the water content exceeds a critical minimum. This minimum water content increases with increasing clay content, increasing the proportion of exchangeable Ca relative to exchangeable Mg, and is influenced by mineralogy (Emerson, 1983). In some soils the presence of organic matter may contribute to a reduction in the critical minimum water content (Emerson, 1983). The critical water content at which mechanical dispersion of clays occurs is close to the plastic limit for some soil materials but is much higher for others (Emerson, 1967). In the former case the desireability of tilling at water contents that create greatest fracturing is offset by the desirability of minimizing the dispersion of clay. The detrimental effects of dispersing clay under these circumstances may, however, be diminished by age hardening if additional stresses are not brought upon the soil immediately after tillage.

Tillage diminishes the continuity of pores within the tilled layer (Ball, 1981) and between the tilled and untilled zones (Goss et al., 1984), can create a compacted zone at the base of the tillage layer (Bowen, 1981), enhances mineralization of organic stabilizing materials (Rovira and Greacen, 1957), disrupts earthworm burrows and increases the susceptibility of earthworms to predation by birds, and can reduce crop residue at the ground surface, thereby enhancing the susceptibility of the surface to erosion. Reducing the extent of tillage will diminish many of these detrimental features of tillage. The suitability of soils for a reduction or elimination of tillage (zero tillage, direct drilling, stubble mulching) relates to the stability and resiliency of the soil. Soils that are least suitable for zero tillage are those which are least stable (particularly with respect to compactive stresses) and are least resilient (particularly with regard to regeneration of macropores) since diminished tillage also diminishes the extent of mechanically formed macropores. Criteria of this nature have been invoked to provide general ratings of the suitability of soils for zero tillage (see Cannell et al., 1978). The seasonal variation in stability provides, however, some latitude in such ratings. For instance, Douglas et al. (1986) have concluded that clay soils may still be suitable for the use of tillage with fall-seeded crops in the United Kingdom, whereas they are not suitable for the production of spring-seeded crops because of the lower stability under spring conditions. The extent to which resiliency is reflected in soil structural form will depend on the time frame and the nature of the stresses still being applied to the soil. For instance, Voorhees and Lindstrom (1984) found that, initially, the introduction of conservation tillage resulted in soil that was less porous at depths of 0 to 15 cm than found under moldboard plowing when measurements were made in a nonwheel-tracked silty clay loam in Minnesota. This difference gradually changed with time, and after 3 to 4 years, conservation tillage produced a higher porosity than did plowing. About 7 years of conservation tillage were needed to produce soil with a porosity equal to that of plowing at a depth of 15 to 30 cm. However, differences in structural characteristics arising from tillage practices were eliminated in the wheel-tracked soil, suggesting that wheel traffic of normal farming operations can negate or mask the regeneration of soil structure arising from conservation tillage.

Management alternatives to minimize the extent of compaction include the use of lighter equipment, increasing the tire width or decreasing the tire pressure, and controlling traffic to traffic lanes. The use of lighter equipment reduces the extent of compaction on a given pass. However, this benefit may be offset somewhat by an increased number of passes and surface area traveled. Increasing the tire width spreads the compactive pressure over a wider area and, while diminishing the amount of compaction of the surface soil immediately under the wheel, appears to do little to diminish subsoil compaction when axle loads are high (Taylor and Burt, 1987). Controlled traffic involves the distinct and permanent separation of cropping zones from tillage lanes (Taylor, 1987). The lanes become compacted, improving tractive efficiency and timeliness of operations. The controlled traffic concept is especially important on those soils which have low stability and resiliency and for those crops in which extensive traffic is unavoidable and the crop value or the length of the growing season makes timeliness of operations very important. Controlled traffic is being used profitably for a range of crops (see Spoor et al., 1988; Tullberg, 1988) and can be easily integrated with reduced tillage when crops are produced on ridges or raised beds.

C. Drainage

Artificial drainage diminishes the length of time that the soil has high water contents and a low structural stability and enhances the extent of wetting and drying. Drainage creates a better soil environment for the growth of plants, soil fauna, and aerobic soil organisms, thereby increasing the rate of addition of carbon to the soil and the rate of turnover of this carbon. There is little information on the rate of change in structural characteristics when artificial drainage is introduced in agricultural soils. Hundal et al. (1976) found that, 16 years after the installation of subsurface drainage in a lacustrine silty clay, the drained soils had higher yields (increase in forage yields of 36%), decreased bulk density (decrease of 5% at 0 to 15 cm in depth), increased volume of air-filled pores between 2- and 100-kPa suction, and higher saturated hydraulic conductivity (increase of 300%) when compared to undrained soil. The undrained soil had the thickest soil crusts and the widest crack spacing under dry conditions.

D. Soil Amendments

Materials are regularly added to soil to alter its physical properties and nutrient levels, to control weeds, insects, or diseases, or simply as a means of disposal. Many of these materials can alter the structural characteristics of soil. Organic amendments selected specifically to improve soil structure include synthetic soil conditioners and selected soil organisms. The potential for synthetic organic amendments in improving the soil structure has been the topic of a series of conferences (deBoodt, 1972; SSSA, 1975; de Boodt and Gabriels, 1976; Emerson et al., 1978) and will not be considered in detail in this chapter. The impact of the different materials on the structural stability varies, and the effectiveness of any one conditioner can vary with soil mineralogy (Page, 1982) and soil conditions at the time of application (Quirk and Williams, 1974; Shaviv et al., 1987). The increase in stability is rapid, although the persistence of the increase with different materials under field conditions is not well documented. The cost of synthetic organic soil-stabilizing materials and their biodegradability (which implies the need for repeated application) continue to constrain widespread use of these materials.

Inoculation of crop residues or soils with organisms that can multiply under natural conditions and which overproduce metabolites that stabilize soil structure may provide an alternative to the repeated application of synthetic stabilizing materials. Preliminary studies involving the inoculation of field plots cropped to corn with a single-cell microalga indicate that this may be a feasible management practice in the future (Metting, 1986).

Animal and poultry manures represent an important organic amendment and repeated applications can lead to increased wet aggregate stability (Tisdall et al., 1978), decreased susceptibility to crust formation (Becher and Kainz, 1983), and increased proportion of large pores (Becher and Kainz, 1983). Improvement in structural characteristics have also been noted from a repeated application of sewage sludge (Pagliai et al., 1981) and other forms of municipal and industrial waste (see Hardan and Al-Ani, 1978; Churchman and Tate, 1986).

Application of biocides can influence the populations of microfauna such as earthworms (Lee, 1985) and soil microflora (Anderson et al., 1981). The impact of these materials on the structural characteristics of soil has largely been ignored.

Inorganic amendments to improve the structural characteristics of soil include materials containing Ca (Shanmuganathan and Oades, 1983b) or, in some cases, compounds of Fe. [Additional details on the evaluation of a range of such compounds are given in Emerson et al. (1978).] These materials have the obvious benefit over organic compounds of not being rendered ineffective over time by biological activity. Their long-term effectiveness would be expected to be determined by rates of leaching losses or crystallization.

Other inorganic soil amendments regularly applied to soils include fertilizer materials. Any effect of fertilization on the soil structure would be expected to be largely a consequence of the effect of fertilizer on the production of roots, the turnover of organic materials in the soil, and to a lesser extent on soil pH. Emmond (1971) observed increases in wet aggregate stability with fertilization except in those treatments where crop residue was removed. However, no effect of fertilization on the structural characteristics of soil were observed in long-term studies by Toogood and Lynch (1959) or Stone et al. (1985). Harris et al. (1966) cite several studies indicating that additional fertilizer diminishes the persistence of aggregation created by adding additional carbon to soil. Potential improvements in structural stability associated with increased production of roots due to fertilization (as contrasted with improvements arising from species selection) may not be fully realized if the fertilization also causes increased rates of mineralization of the stabilizing materials. These effects will be accentuated if additional fertilizer results in a reduction in the production of mucilages. Elliott and Lynch (1984) observed significantly greater aggregation on the addition of wheat straw with low N contents (0.25% N w/w) to volcanic ash than was observed when straw with higher N contents (up to 1.09% N w/w) was added. The differences were attributed to the cessation of biomass production, which has been observed (Knapp et al., 1983) when N becomes limiting during the utilization of readily available carbon pools and the subsequent synthesis of the remainder of the available carbon into polysaccharides by the cells.

V. Quantifying the Influence of Cropping Systems on the Rates of Change of Soil Structure

Changes in cropping practices, tillage, traffic, drainage, and the use of soil amendments alter the rate of change of the soil structure, but it is not possible at present to predict quantitatively how much a change in cropping systems will alter the rate of change of the most important soil structural characteristics. Yet for reasons that have already been cited, such information would be useful to farmers and advisory personnel. Our inability to make such predictions is due, in part, to the large number of possible combinations of crops and management practices coupled with the large number of structural characteristics and the variations in their relative importance with soil, climate, and crops. The challenge facing researchers is to identify the least data that are necessary to permit extrapolations from several cropping systems to other cropping systems on a single soil or to extrapolate the response of a series of soils to the introduction of new cropping systems to other soils. Identification of the minimum data set would be facilitated by the use of models to describe the influence of cropping systems on the rates of change of soil structure. Determination of the rate constants for a range of soils and management practices would make it possible to predict rate constants from inherent soil properties, characteristics of the crop grown, and specific management practices. Predictions of rates of change of the structural characteristics of different soils under different crops and different management practices would then follow. In addition to operational uses, such models could provide an incentive to integrate in a systematic way many of the processes that influence soil structure, thereby leading to an improved understanding of the dynamics of changes of structure.

A. Conceptual Models

A model to quantify the influence of cropping systems on the structural characteristics of soils must, at this stage, be conceptual in nature. Boone (1986) has provided a simplified block diagram showing the relation between some soil properties, climate, and plant behavior when soil compaction occurs. Gibbs and Reid (1988) have emphasized the importance of macropores and have developed a form of budget equation to describe the change in the volume fraction of pores greater than 100 μm in diameter, ΔE_{100}, over a given period,

$$\Delta E_{100} = T + B_e + R_d - R_g - S \qquad (2)$$

where T is the net change due to tillage, B_e is the change due to earthworm burrowing, R_d is the change due to the decomposition of roots, R_g is the change due to root growth, and S is the change in macroporosity due to slaking, slumping, and irreversible shrinkage. The authors note that characterizing the effect of a system on macroporosity with greatest accuracy might need the determination of the variables in Equation (2) for several short consecutive periods, especially if tillage is intermittent and if S can be expected to be rapid. This model provides an elegant and relatively simple approach to quantifying the influence of cropping systems on an important characteristic of the structural form. The model will be most applicable to soils and cropping systems in which the most important limitations to growth relate to macropores and in which the concentration of structure-stabilizing materials does not change with time.

A conceptual model that is applicable to any characteristics of structural form, and which attempts to link the temporal variation in structural form with stability and the level of stabilizing materials, has been developed by Kay et al. (1988). The variation with time t of a characteristic S of the structural form for a given cropping system relative to the value of the corresponding characteristic S_0 of the soil in a reference state is given by

$$\frac{S}{S_0} = f(t) \tag{3}$$

Soil in the reference state must be part of an ecosystem that is stable within the time scale of interest. The time scale would be one of years rather than decades and would be compatible with the longest term a given cropping system is likely to be practiced. The time scale would also have to be long enough that variations within seasons that do not contribute to longer term trends can be distinguished from the longer term trends.

The first hypothesis upon which the model is based is that the rate of change of a relative value of a characteristic of structural form can be related to two functions that are additive, a biological function and a physical-chemical function. The biological function B describes the rate of change in a characteristic of the structural form caused by living organisms. Examples include the alteration of total porosity by earthworms and other soil fauna. The biological function only includes the effect of organisms on the structural form; the influence of organisms on the stabilization of the structure is considered in the physical-chemical function. The physical-chemical function includes two variables, the force per unit area F/A applied to the soil (both cultivated and reference conditions) and a relative stability parameter $(R/R_0)_{S,F}$ which is appropriate for a given characteristic of structural form and a given stress. Stresses include those that may change with cropping systems, such as tensile, shear, and compressive stresses applied by tillage, traffic, or root development, and osmotic stresses. Stresses that are less sensitive to cropping systems, such as stresses associated with rapid wetting, drying, and freezing, are also included in the physical-chemical function. The biological function and the physical-chemical function would be related to the rate of change in structural form as follows:

$$\frac{\partial S/S_0}{\partial t} = f\left(\frac{R}{R_0}, \frac{F}{A}\right) + f\left(\frac{B}{B_0}\right) \tag{4}$$

The forces included in the physical-chemical function can be both regenerative and destructive, and their individual impact on the structural form will depend on the magnitude of the stress and the magnitude of the stability parameter. The summation of regenerative and destructive processes will determine whether the physical-chemical function makes a negative or a positive contribution to the rate of change of a specific characteristic of the structural form. Soils which are structurally very resilient will experience a rapid regeneration of structural form when forces of a destructive mechanical nature are removed.

The relation between stress, stability, and structural form is well established for some parameters (such as, applied pressure, compression index, and porosity). However, relations involving other parameters are not as well defined, and such relations must be defined before a number of stability parameters can be used in Equation (4).

The second hypothesis upon which the model is based is that progressive changes in relative stability arising from the introduction of a cropping system are due to a progressive change in the relative concentration of stabilizing materials (C/C_0), that is,

$$\left(\frac{R}{R_0}\right)_{S,F} = f\left(\frac{C}{C_0}\right)_{S,F} \tag{5}$$

$$\left(\frac{C}{C_0}\right)_{S,F} = f(t) \tag{6}$$

Different stabilizing materials may be important when different stability characteristics are considered (Tisdall and Oades, 1982), and the stability characteristics are specific for each combination of structural form and stress. Only those stabilizing materials which, by virtue of their association with failure zones, will be sensitive to changes in cropping systems would therefore be included in Equations (5) and (6). Changes in the concentration of these materials may arise from changes in soil pH, salinity, aeration, the extent of leaching, application of soil amendments, and the amount and distribution of photosynthate deposited below ground. Equation (6) would be expected to be specific for a given soil. If information is not available on either the identity of the stabilizing materials relevant to a specific stability characteristic or the temporal variation in the concentration of the material under a given cropping system, Equation (6) can be substituted in Equation (5) to give

$$\left(\frac{R}{R_0}\right)_{S,F} = f(t) \tag{7}$$

The hypothesis that the relative stability is a function of the relative level of stabilizing materials is an obvious simplification. Cropping systems that alter physical processes influencing the strength of failure zones (such as an alteration in the extent of wetting and drying) may alter the stabilizing materials and therefore their "effective" concentration, but may also alter the strength of failure zones in ways not involving stabilizing materials (such as the formation of microcracks). Such processes would be hidden in Equation (7) when it is used.

The use of relative values of structural form, stability, concentration of stabilizing materials, and biological activity, facilitate an evaluation of different cropping systems on a soil in terms of movement toward or away from the reference state as well as a comparison of responses between soils to a given cropping system. The use of relative values also minimizes variations in each of the parameters due to environmental factors, such as temperature and moisture content, provided the influence of these factors on the structural parameters is similar in both the reference and the cropped state. In those cases in which cropped and reference states respond differently to environmental factors or are exposed to much different stresses, it may be necessary to use absolute values of parameters in the equations rather than relative values.

B. Form of Functions in Models

The form of specific functions in the conceptual models must be defined and the constants in the functions quantified before the conceptual models can be used. Boone (1986), Gibbs and Reid (1988), and Kay et al. (1988) have discussed the functions that are required. Rather than discuss all of the different functions in this chapter, a specific function will be selected and the form of the function illustrated using existing data. In the process the value of conceptual models in defining the nature of data that should be collected and the insights gained with regard to changes in soil structure will become apparent.

The relation selected for examination is the temporal change in wet aggregate stability (Equation 7) when grass is introduced on land that has been used to produce arable crops for an extended period of time. This relation was selected largely because of the availability of useful data. Low (1955) characterized the increase in the stability of aggregates greater than 2 mm in diameter at a depth of 0 to 15 cm in predominantly clay soils in the United Kingdom when old arable land was sown to perennial ryegrass and white clover. Corresponding data for a permanent pasture (about 100 years old) at each site were also collected and have been used as reference state in recalculating the data. Greacen (1958) measured the increase in stability of aggregates greater than 2 mm in diameter on a fine sandy loam in Australia subsequent to the introduction of grass and subterranean clover. Stabilities were measured at different depths, and the data have been recalculated to give an average stability at a depth of 0 to 15 cm. Corresponding data on aggregate stability were also obtained by sampling a virgin soil and are used as reference values. Jastrow (1987) obtained data on the stability of different aggregate size fractions at 0 to 10 cm depth during the revegetation of restored prairie on a mixed silt to silty loam in Illinois, United States. Stabilities of aggregates >250 μm were selected and data from a virgin area of remnant prairie were used as reference data. Perfect et al. (1989b) obtained data on the increase in stability of aggregates >250 μm diameter at 0 to 15 cm depth where pure strands of different forage species were introduced on a silt loam soil in Ontario, Canada. Regression equations were developed to describe the temporal variations in stability, and the equations included a maximum stability that was used as the reference state. Data for their high nitrogen bromegrass treatment are used in this comparison. The temporal variations in the relative wet aggregate stability from the four studies are shown in Figure 8 and indicate similar shapes of functions in spite of the differences in soils, climate, and crop species, as well as differences in methods of determining the stability.

The regression equation found by Perfect et al. (1989b) to best describe their data was of the form

$$\mathrm{WAS} = \mathrm{WAS}_{\max} - Me^{bt} \tag{8}$$

where WAS and WAS_{max} are the measured and maximum wet aggregate stabilities, respectively, M is a constant which is a function of water content, b is the rate constant, and t is time in years.

If at $t = 0$, WAS = $WAS_{t=0}$, then Equation (8) can be rewritten as

$$\frac{WAS_{max} - WAS}{WAS_{max} - WAS_{t=0}} = e^{bt} \tag{9}$$

This equation "normalizes" the change in wet aggregate stability with respect to the maximum possible change. A plot of ln $[(WAS_{max} - WAS)/(WAS_{max} - WAS_{t=0})]$ versus t gives a straight line whose slope is the rate constant. Data from Figure 8 are replotted following Equation (9) in Figure 9. The same function appears to be applicable under a broad range of soils, crops, and climatic conditions. The range in rate constants in Figure 9 (0.08 to 0.30 per year) is of the same order as the range in rate constants Perfect et al. (1989b) observed for different forage species on the same soil (0.10 per year for red clover to 0.15 per year for bromegrass), which is rather surprising given the differences in soils and climates.

The rate constant can be used to characterize the time required ($t_{0.5}$) for a forage crop to improve the structural stability to a point midway between the initial level and maximum stability (Perfect et al., 1989b), that is, $t_{0.5} = \ln 0.5/b$. Values of $t_{0.5}$ can be tabulated to compare the effects of different crops on the wet aggregate stability of a soil or to compare the effects of a crop on the wet aggregate stability of a range of soils. Values of $t_{0.5}$ provide a convenient basis for extrapolating from situations where the response has been measured to those cases where data are not available. If the relative wet aggregate stability is related

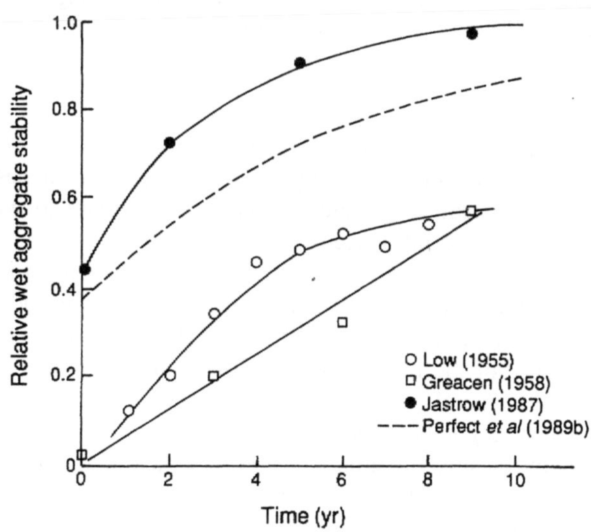

Figure 8. Increase in wet aggregate stability subsequent to planting forages on arable land (Equation 7).

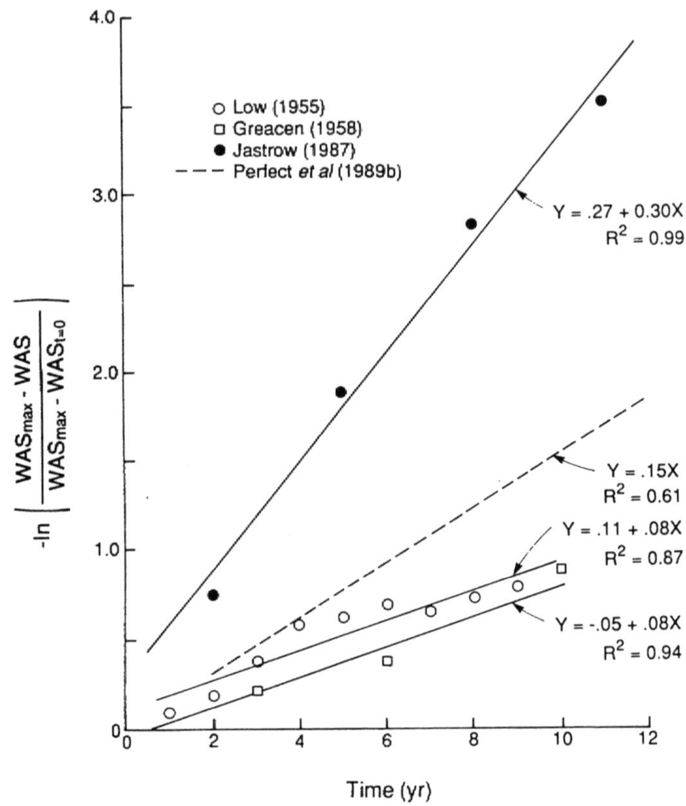

Figure 9. Increase in wet aggregate stability subsequent to planting forages on arable land, Equation (9). Slopes of lines are rate constants.

linearly to the relative concentration of stabilizing material, it can be shown that the rate constant in Equation (9) is equal to the rate constant describing the accumulation of the stabilizing materials that are important to wet aggretate stability when forages are introduced on old arable land.

Additional studies are necessary to determine whether temporal changes in other stability parameters can be described by the same function and, if they can, the extent to which the rate constants differ for different stability characteristics and the dependence of the rate constants on inherent soil properties.

The analysis of temporal variation in wet aggregate stability suggests that data on stability which might be collected in the future will be most useful for extrapolation between sites if the initial stability is measured, and if the relative stability is measured over either long terms (as done by Low, 1955; Greacen, 1958; or Jastrow, 1987) or short terms, with the facility of removing seasonal variations in the relative stability and estimating the maximum stability (as done by Perfect et al., 1989b).

VI. Summary

Soil structure can change in response to different agricultural practices and can, in turn, influence both the productivity arising from agricultural practices and the impact of these practices on the environment. Soil structure can be characterized in terms of form, stability, and resiliency. A number of characteristics are used to describe these three dimensions of soil structure, and the relative importance of these characteristics vis-à-vis productivity or environmental quality will vary with soil, climate, and crop.

Cropping systems which sustain soil productivity and minimize deterioration of the environment continue to be a major goal of agricultural research programs. Information on the rates of change in soil structural characteristics caused by existing and new cropping systems under different soil and climatic conditions is essential to meeting this goal. The tremendous number of combinations of cropping systems, soils, climate, and structural characteristics makes it impractical to try to collect data on all combinations. It is therefore essential to have the capability to make accurate predictions of rates of change of soil structure for the multitude of cases where data will not exist. This capability will be enhanced by models that relate rates of the change of structural characteristics of soils to those physical, chemical, and biological processes that occur in soil and are influenced by management practices. Major challenges facing researchers in the development of these models will include relating stability measurements to parameters of structural form (thereby linking a vast wealth of information on the dynamics of structure-stabilizing materials with information on the persistence of characteristics of soil pores, failure zones, and aggregates). In addition, models that relate crop yield to the structural characteristics of soil for a range of climatic conditions and crops are needed in order to make information on rates of change in soil structure of greatest use at a farm level.

Acknowledgments

This chapter was prepared while the author was on sabbatical leave in the Department of Soil Science, Waite Agricultural Research Institute, University of Adelaide, Australia. Financial support for travel was provided by the University of Adelaide and the Natural Sciences and Engineering Research Council of Canada and is gratefully acknowledged. Special appreciation is expressed to members of the Department of Soil Science who contributed to this chapter through numerous discussions with the author and who contributed constructive criticisms of earlier drafts. Foremost among this group are A. R. Dexter, J. M. Oades, J. A. Baldock, C. D. Grant, and J. Marinissen.

References

Abu-Sharar, T. M., F. T. Bingham, and J. D. Rhoades, 1987. Reduction in hydraulic conductivity in relation to clay dispersion and disaggregation. *Soil Sci. Soc. Amer. J.* 51:342–346.

Adu, J. K., and J. M. Oades. 1978. Physical factors influencing decomposition of organic materials in soil aggregates. *Soil Biol. Biochem.* 10:109–115.

Alderfer, R. B. 1946. Seasonal variability in the aggregation of Hagerstown silt loam. *Soil Sci.* 62:151–168.

Ali, O. M., M. Yousaf, and J. D. Rhoades. 1987. Effect of exchangeable cation and electrolyte concentration on mineralogy of clay dispersed from aggregates. *Soil Sci. Soc. Amer. J.* 51:896–900.

Anderson, J. P. E., R. A. Armstrong, and S. N. Smith. 1981. Methods to evaluate pesticide damage to the biomass of the soil microflora. *Soil Biol. Biochem.* 13:149–153.

Angers, D. A., B. D. Kay, and P. H. Groenevelt. 1987. Compaction characteristics of a soil cropped to corn and bromegrass. *Soil Sci. Soc. Amer. J.* 51:779–783.

Arya, L. M., and G. R. Blake. 1972. Stabilization of newly formed soil aggregates. *Agron. J.* 64:177–180.

Baldock, J. A., and B. D. Kay. 1987. Influence of cropping history and chemical treatments on the water stable aggregation of a silt loam soil. *Canad. J. Soil Sci.* 67:501–511.

Ball, B. C. 1981. Pore characteristics of soils from two cultivation experiments as shown by gas diffusivities and permeabilities and air-filled porosities. *J. Soil Sci.* 32:483–498.

Barber, D. A., and J. K. Martin. 1976. The release of organic substances by cereal roots into soil. *New Phytologist* 76:69–80.

Becher, H. H., and M. Kainz. 1983. [Effects of long-term manuring on soil structure]. *Z. Acker- Pflanzenbau* 152:152–158.

Bennett, O. L., E. L. Mathias, and C. B. Sperow. 1976. Double cropping for hay and no-tillage corn production as affected by sod species with rates of atrazine and nitrogen. *Agron. J.* 68:250–254.

Benoit, G. R. 1973. Effect of freeze–thaw cycles on aggregate stability and hydraulic conductivity of three aggregate sizes. *Soil Sci. Soc. Amer. Proc.* 37:3–5.

Bisal, F., and W. S. Ferguson. 1968. Monthly and yearly changes in aggregate size of surface soils. *Canad. J. Soil Sci.* 48:159–164.

Bisal, F., and K. F. Nielsen. 1964. Soil aggregates do not necessarily break down over winter. *Soil Sci.* 98:345–346.

Blackmore, A. V. 1981. Self-mulching soils. *Soils News* (Aust. Soc. Soil Sci.) July, p. 4.

Blake, G. R., W. W. Nelson, and R. R. Allmaras. 1976. Persistence of subsoil compaction in a Mollisol. *Soil Sci. Soc. Amer. J.* 40:943–948.

Boekel, P. 1965. Handhaving van een goede bodemstructuur op klei-en zavelgronden. *Landbouwk. Tijdschr.* 77:842–849.

Boone, F. R. 1986. Towards soil compaction limits for crop growth. *Neth. J. Agric. Sci.* 34:349–360.

Bosatta, E., and G. L. Agren. 1985. Theoretical analysis of decomposition of heterogeneous substrates. *Soil Biol. Biochem.* 17:601–610.

Bowen, H. D. 1981. Alleviating mechanical impedance in modifying the root environment to reduce stress. In: G. F. Arkin and H. M. Taylor (eds.), ASAE Mono. 4, Amer. Soc. Agr. Eng., St Joseph, MI.

Braunack, M. V., J. S. Hewitt, and A. R. Dexter. 1979. Brittle fracture of soil aggregates and the compaction of aggregate beds. *J. Soil Sci.* 30:653–667.

Brewer, R., and A. D. Haldane. 1957. Preliminary experiments in the development of clay orientation in soil. *Soil Sci.* 84:301–309.

Bui Huu Tri. 1968. Dynamique de la granulation du sol sous prairie. *Ann. Agron.* 19:415–439.

Bullock, M. S., W. D. Kemper, and S. D. Nelson. 1988. Soil cohesion as affected by freezing, water content, time and tillage. *Soil Sci. Soc. Amer. J.* 52:770–776.

Bullock, P., A. C. D. Newman, and A. J. Thomasson. 1985. Porosity aspects of the regeneration of soil structure after compaction. *Soil Tillage Res.* 5:325–341.

Burns, R. G., and J. A. Davies. 1986. The microbology of soil structure. *Biol. Agric. Hort.* 3:95–113.

Cannell, R. Q., D. B. Davies, D. Mackney, and J. D. Pidgeon. 1978. The suitability of soils for sequential direct drilling of combined harvested crops in Britain: a provisional classification. *Outlook on Agric.* 9:306–316.

Cassel, D. K. 1983. Spatial and temporal variability in soil physical properties following tillage of Norfolk loamy sand. *Soil Sci. Soc. Amer. J.* 47:196–201.

Chamberlain, E. J., and A. J. Gow. 1979. Effect of freezing and thawing on the permeability and structure of soils. *Eng. Geol.* 13:73–92.

Chan, K. Y. 1989a. Friability of a hardsetting soil under different tillage and land use practices. *Soil Tillage Res.* 13:287–298.

Chan, K. Y. 1989b. Effect of tillage on aggregate strength and aggregation of vertisols. *Soil Tillage Res.* 13:163–175.

Churchman, G. J., and K. R. Tate. 1986. Effect of slaughterhouse effluent and water irrigation upon aggregation in seasonally dry New Zealand soil under pasture. *Aust. J. Soil Res.* 24:505–516.

Churchman, G. J., and K. R. Tate. 1987. Stability of aggregates of different size grades in allophanic soils from volcanic ash in New Zealand. *J. Soil Sci.* 38:19–27.

Coote, D., C. A. Malcolm-McGovern, G. J. Wall, W. T. Dickinson, and R. P. Rudra. 1988. Seasonal variation of erodibility indices based on shear strength and aggregate stability in some Ontario soils. *Canad. J. Soil Sci.* 68:405–416.

Coughlan, K. J., W. E. Fox, and J. D. Hughes. 1973. Aggregation in swelling clay soils. *Aust. J. Soil Res.* 11:133–141.

Croney, D., and J. D. Coleman. 1954. Soil structure in relation to soil suction (pF). *J. Soil Sci.* 5:75–84.

Croney, D., J. D. Coleman, and W. P. M. Black. 1958. The movement and distribution of water in soil in relation to highway design and performance. Spec. Rep. 40, Highway Res. Bd., Washington, DC, pp. 226–252.

Czeratzki, W., and H. Frese. 1958. Importance of water in formation of soil structure. Spec. Rep. 40, Highway Res. Bd., Washington, DC, pp. 200–211.

Dasog, G. S., D. F. Acton, A. R. Mermut, and E. de Jong. 1988. Shrink swell potential and cracking in clay soils of Saskatchewan. *Canad. J. Soil Sci.* 68:251–260.

Davenport, J. R., and R. L. Thomas. 1988. Carbon partitioning and rhizodeposition in corn and bromegrass. *Canad. J. Soil Sci.* 68:693–701.

Davenport, J. R., R. L. Thomas, and S. C. Mott. 1988. Carbon mineralization of corn (*Zea mays* L.) and bromegrass (*Bromus inermis* Leyss.) components with an emphasis on the below-ground carbon. *Soil Biol. Biochem.* 20:471–476.

Davies, P. 1985. Influence of organic matter content, moisture status and time after reworking on soil shear strength. *J. Soil Sci.* 36:299–306.

de Boodt, M. (ed.) 1972. *Fundamentals of soil conditioning*. Meded. 37, Fak. Landbouwwet., Rijksuniversiteit, Gent.

de Boodt, M., and D. Gabriels (eds). 1976. *Proc. 3rd Int. Symp. on Soil Conditioning*. Meded., Fak. Landbouwwet. Rijksuniversiteit. Gent.

Dexter, A. R. 1985. Shapes of aggregates from tilled layers of some Dutch and Australian soils. *Geoderma* 35:91–107.

Dexter, A. R. 1986a. Model experiments on the behaviour of roots at the interface between a tilled seed-bed and a compacted sub-soil II: Entry of pea and wheat roots into sub-soil cracks. *Plant and Soil* 95:135–147.

Dexter, A. R. 1986b. Model experiments on the behaviour of rots at the interface between a tilled seed-bed and a compacted sub-soil. *Plant and Soil* 95:149–161.

Dexter, A. R. 1987a. Mechanics of root growth. *Plant and Soil* 98:303–312.

Dexter, A. R. 1987b. Compression of soil around roots. *Plant and Soil* 97:401–406.

Dexter, A. R. 1988. Advances in characterization of soil structure. *Soil Tillage Res.* 11:199–238.

Dexter, A. R., R. Horn, and W. D. Kemper. 1988. Two mechanisms for age-hardening of soil. *J. Soil Sci.* 39:163–175.

Dexter, A. R., B. Kroesbergen, and H. Kuipers. 1984. Some mechanical properties of aggregates of top soils from the IJsselmeer polders. 2: Remolded soil aggregates and the effect of wetting and drying cycles. *Neth. J. Agric. Sci.* 32:215–227.

Donald, R. G., B. D. Kay, and M. H. Miller. 1987. The effect of soil aggregate size on early shoot and root growth of maize (*Zea mays* L.). *Plant Soil* 103:251–259.

Doran, J. W., L. N. Mielke, and S. Stamatiadas. 1988. Microbial activity and N cycling as regulated by soil water-filled pore space. pp. 49–54. In: Proc. 11th Int. Conf., Int. Soil Tillage Res. Org., Edinburgh, Scotland, vol. 1.

Douglas, J. T. 1986. Effects of season and management on the vane shear strength of a clay topsoil. *J. Soil Sci.* 37:669–679.

Douglas, J. T., M. G. Jarvis, K. R. House, and M. J. Goss. 1986. Structure of a silty loam in relation to management. *J. Soil Sci.* 37:137–151.

Duffey, J. E., H. Halen, and R. Frankart. 1986. Evolution de la stabilité structurale du sol sous l'influence des racines de trèfle (*Trifolium pratense* L.) et de ray-grass (*Lolium multiflorum* Lmk): Observations pendant et après culture. *Agronomie* 6:811–817.

Dwyer, L. M., D. W. Stewart, and D. Balchin. 1988. Rooting characteristics of corn soybeans and barley as a function of available water and soil physical characteristics. *Canad. J. Soil Sci.* 68:121–132.

Edwards, C. A., and J. R. Lofty. 1972. *Biology of earthworms*. Chapman and Hall, London.

Edwards, W. M., L. D. Norton, and C. E. Redmond. 1988. Characterizing macropores that affect infltration into nontilled soil. *Soil Sci. Soc. Amer. J.* 52:483–487.

Egashira, K., Y. Kaetsu, and K. Takkuma. 1983. Aggregate stability as an index of erodibility of Ando soils. *Soil Sci. Plant Nutr.* 29:473–481.

Elkins, C. B., R. L. Haaland, and C. S. Hoveland. 1977. Grass roots as a tool for penetrating soil hardpans and increasing crop yields. pp. 21–26. In: *Proc. 34th Southern Pasture and Forage Improvement Conf.*, Auburn AL.

Elliott, E. T. 1986. Aggregate structure and carbon, nitrogen, and phosphorus in native and cultivated soils. *Soil Sci. Soc. Amer. J.* 50:627–633.

Elliott, L. F., and J. M. Lynch. 1984. The effect of available carbon and nitrogen in straw on soil and ash aggregation and acetic acid production. *Plant and Soil* 78:335–343.

Elliott, E. T., R. V. Anderson, D. C. Coleman, and C. V. Cole. 1980. Habitable pore space and microbial trophic interactions. *Oikos* 34:327–335.

Emerson, W. W. 1954. The determination of the stability of soil crumbs. *J. Soil Sci.* 5:235–350.

Emerson, W. W. 1955. The rate of water uptake of soil crumbs at low suctions. *J. Soil Sci.* 6:147–159.

Emerson, W. W. 1959. The structure of soil crumbs. *J. Soil Sci.* 10:235–243.

Emerson, W. W. 1967. A classification of soil aggregates based on their coherence in water. *Aust. J. Soil Res.* 5:47–57.

Emerson, W. W. 1983. Inter-particle bonding in soils: An Australian viewpoint. Div. of Soils, CSIRO, Melbourne. Academic Press, London, pp. 477–498.

Emerson, W. W., R. D. Bond, and A. R. Dexter. 1978. *Modification of soil structure*. Wiley, Chichester, England.

Emerson, W. W., R. C. Foster, and J. M. Oades. 1986. Organo-mineral complexes in relation to soil aggregation and structure. In: *Interaction of soil minerals with natural organics and microbes*. SSSA Spec. Pub. 17, Soil Sci. Soc. Amer., Madison WI.

Emmond, G. S. 1971. Effect of rotations, tillage treatments and fertilizers on the aggregation of a clay soil. *Canad. J. Soil Sci.* 51:235–241.

Foster, R. C. 1988. Microenvironments of soil microorganisms. *Biol. Fertil. Soils* 6:189–203.

Fox, W. E. 1964. Cracking characteristics and field capacity in a swelling soil. *Soil Sci.* 98:413.

Francis, P. B., and R. M. Cruse. 1983. Soil water matric potential effects on aggregate stability. *Soil Sci. Soc. Amer. J.* 47:578–581.

Gameda, S., G. S. V. Raghaven, E. McKeyes, and R. Theriault. 1987. Subsoil compaction in a clay soil. I: Cumulative effects. *Soil Tillage Res.* 10:113–122.

Gerard, C. J. 1987. Laboratory experiments on the effects of antecedent moisture and residue application on aggregation of different soils. *Soil Tillage Res.* 9:21–32.

Gibbs, R. J., and J. B. Reid. 1988. A conceptual model of changes in soil structure under different cropping systems. *Adv. Soil Sci.* 8:123–149.

Gifford, R. O., and D. F. Thran. 1974. Bonding mechanisms for soil crusts: Part II. Strength of silica cementation. In: J. W. Cary and D. D. Evans (eds.), *Soil crusts*. Tech. Bull. 214, Agric. Res. Stn., Univ. Ariz., Tucson.

Goss, M. J., W. Ehlers, F. R. Boone, I. White, and K. R. House. 1984. Effects of soil management practice on soil physical conditions affecting root growth. *J. Agric. Eng. Res.* 30:131–140.

Grant, C. D. 1989. Soil structure and tensile strength in relation to the microtopography of soil fracture surfaces. Unpublished Ph.D. thesis, Dept. Soil Sci., Waite Agric. Res. Inst., Univ. of Adelaide, Adelaide, Australia.

Grant, C. D., and A. R. Dexter. 1989. Generation of microcracks in moulded soils by rapid wetting. *Aust. J. Soil Res.* 27:169–182.

Grant, C. D., B. D. Kay, P. H. Groenevelt, G. E. Kidd, and G. W. Thurtell. 1985. Spectral analysis of micropenetrometer data to characterize soil structure. *Canad. J. Soil Sci.* 65:789–804.

Greacen, E. L. 1958. The soil structure under pastures. *Aust. J. Soil Res.* 9:129–137.

Greacen, E. L., D. A. Farrell, and B. Cockroft. 1968. Soil resistance to metal probes and plant roots. *Trans. 9th Int. Cong. Soil Sci., Adelaide*, vol. 1. Angus and Robertson, Sydney, Australia, pp. 769–779.

Greenland, D. J., D. Rimmer, and D. Payne. 1975. Determination of the structural stability class of English and Welsh soils, using a water coherence test. *J. Soil Sci.* 26:294–303.

Gregorich, E., R. G. Kachanoski, and R. P. Voroney. 1988. Ultrasonic dispersion of aggregates: Distribution of organic matter in size fractions. *Canad. J. Soil Sci.* 68:395–403.

Grevers, M. C. G., and E. de Jong. 1988. The use of differential imagery in the characterization of soil macrostructure. In: *Abstracts, Canad. Soc. Soil Sci. Ann. Meet.*, Univ. of Calgary, Alberta, Canada.

Groenevelt, P. H., and B. D. Kay. 1981. On pressure distribution and effective stress in unsaturated soils. *Canad. J. Soil Sci.* 61:431–443.

Groenevelt, P. H., B. D. Kay, and C. D. Grant. 1984. Physical assessment of a soil with respect to rooting potential. *Geoderma* 34:101–114.

Guerif, J. 1988. Résistance en traction des agrégats terreus: Influence de la texture de la matière organique et de la teneur en eau. *Agronomie* 8:379–386.

Guidi, G., G. Poggio, and G. Petruzelli. 1985. The porosity of soil aggregates from bulk soil and soil adhering to roots. *Plant Soil* 87:311–314.

Gumbel, E. J. 1960. *Statistics of extremes*. Columbia Univ. Press, NY.

Hadas, A. 1987a. Dependence of "true" surface energy of soils on air entry pore size and chemical constituents. *Soil Sci. Soc. Amer. J.* 51:187–191.

Hadas, A. 1987b. Long-term tillage practice effects on soil aggregation modes and strength. *Soil Sci. Soc. Amer. J.* 51:191–197.

Hakansson, I. 1985. Swedish experiments on sub-soil compaction by vehicles with high axle load. *Soil Use Mgt.* 1:113–116.

Hakansson, I., W. B. Voorhees, and H. Riley. 1988. Vehicle and wheel factors influencing soil compaction and crop response in different traffic regimes. *Soil Tillage Res.* 11:239–282.

Hamblin, A. 1985. The influence of soil structure on water movement, crop growth and water uptake. *Adv. Agron.* 38:95–155.

Hantschel, R., M. Kaupenjohann, R. Horn, J. Gradl, and W. Zech. 1988. Ecologically important differences between equilibrium and percolation soil extracts, Bavaria. *Geoderma* 43:213–227.

Hardan, A., and A. N. Al-Ani. 1978. Improvement of soil structure by using date and sugar beet waste products. In: W. W. Emerson, R. D. J. Bond and A. R. Dexter (eds.), *Modification of soil structure*. Wiley, Chichester, England.

Harris, R. F., G. Chesters, and O. N. Allen. 1966. Dynamics of soil aggregation. *Adv. Agron.* 18:107–169.

Heinonen, R. 1977. Towards a "normal" soil bulk density. *Soil Sci. Soc. Amer. J.* 41:1214–1215.

Heinonen, R. 1982. Alleviation of soil compaction by natural forces and cultural practices. In: *Proc. Int. Conf. Land Clearing and Development,* IITA, Ibadan, Nigeria.

Helal, H. M., and D. R. Sauerbeck. 1984. Influence of plant roots on C and P metabolism in soil. *Plant Soil* 76:175–182.

Hetier, J. M., F. Andreux, E. Schouller, and C. Marol. 1986. Organic matter inputs to soil after growth of carbon-14-nitrogen-15 labelled maize. *Soil Sci. Soc. Amer. J.* 50:76–80.

Hettiaratchi, D. 1988. Theoretical soil mechanics and implement design. *Soil Tillage Res.* 11:325–348.

Horn, R., and A. R. Dexter. 1989. Dynamics of soil aggregation in an irrigated desert loess. *Soil Tillage Res.* 13:253–266.

Horn, R., P. S. Blackwell, and R. White. 1988. The effect of stress duration on the pressure transmission in an ameliorated red-brown earth under irrigation. In: *Proc 11th Int. Conf.*, Int. Soil and Tillage Res. Org., Edinburgh, Scotland, vol. 2, pp. 492–498.

Hundal, S. S., G. O. Schwab, and G. S. Taylor. 1976. Drainage system effects on physical properties of a lakebed clay soil. *Soil Sci. Soc. Amer. J.* 40:300–305.

Ingles, O. G. 1962. Bonding forces in soils. Part III: A theory of tensile strength for stabilized and naturally coherent soils. *Proc. Aust. Rd. Res. Bd.* 1(2):1025–1047.

Jastrow, J. D. 1987. Changes in soil aggregation associated with tallgrass prairie restoration. *Amer. J. Bot.* 74:1656–1664.

Jenkinson, D. S., and J. H. Rayner. 1977. The turnover of soil organic matter in some of the Rothamsted classical experiments. *Soil Sci.* 123:298–305.

Johnston, A. E. 1986. Soil organic matter, effects on soils and crops. *Soil Use Mgt.* 2:97–105.

Katou, H., K. Miyaji, and T. Kubota. 1987. Susceptibility of undisturbed soils to compression as evaluated from the changes in the soil water characteristic curves. *Soil Sci. Plant Nutr.* 33:539–554.

Kay, B. D., and E. Perfect. 1988. State of the art: heat and mass transfer in freezing soils. In: *Proc. 5th Int. Symp. Ground Freezing,* Nottingham, UK, vol. 1, pp. 3–22.

Kay, B. D., C. D. Grant, and P. H. Groenevelt. 1985. Significance of ground freezing on soil bulk density under zero tillage. *Soil Sci. Soc. Amer. J.* 49:973–978.

Kay, B. D., D. A. Angers, P. H. Groenevelt, and J. A. Baldock. 1988. Quantifying the influence of cropping history on soil structure. *Canad. J. Soil Sci.* 68:359–368.

Kemper, W. D., and R. C. Rosenau. 1984. Soil cohesion as affected by time and water content. *Soil Sci. Soc. Amer. J.* 48:1001–1006.

Kemper, W. D., R. C. Rosenau, and A. R. Dexter. 1987. Cohesion development in disrupted soils as affected by clay and organic matter content and temperature. *Soil Sci. Soc. Amer. J.* 51:860–867.

Kemper, W. D., M. Bullock, and A. R. Dexter. 1988. Soil cohesion changes. pp. 81–95. In: W. E. Larson et al. (eds.), *Mechanics and related processes in structured agricultural soils.* Kluwer Acad. Publ., Boston, MA.

Kirby, P. C., and G. R. Mehuys. 1987. Seasonal variation of soil erodibilities in southwestern Quebec. *J. Soil Water Conserv.* 42:211–215.

Kladivko, E. J., A. D. Mackay, and J. M. Bradford. 1986. Earthworms as a factor in the reduction of soil crusting. *Soil Sci. Soc. Amer. J.* 50:191–196.

Knapp, E. B., L. F. Elliott, and G. S. Campbell. 1983. Carbon, nitrogen and microbial biomas interrelationships during the decomposition of wheat straw: a mechanistic simulation model. *Soil Biol. Biochem.* 15:455–461.

Koolen, A. J., and H. Kuipers. 1983. *Agricultural soil mechanics.* Springer-Verlag, NY.

Kretschmer, A. 1982. Description des galerres des vers de terre et variation saisonnière des reseaux (observation en conditions naturelles). *Rev. Ecol. Biol. Sol.* 19:579–591.

Lal, P., G. F. Wilson, and B. N. Okigbo. 1979. Changes in properties of an alfisol produced by various cover crops. *Soil Sci.* 127:377–382.

Ladd, C. C. 1960. Mechanisms of swelling by compacted clay. Bull. 245, Highway Res. Bd., Washington, DC, pp. 10–26.

Larson, W. E., S. C. Gupta, and R. A. Useche. 1980. Compression of agricultural soils from eight soil orders. *Soil Sci. Soc. Amer. J.* 44:450–457.

Lee, K. E. 1985. *Earthworms: Their ecology and relationships with soils and land use.* Academic Press, Sydney, Australia.

Letey, J. 1985. Relationship between soil physical properties and crop productions. pp. 277–294. In: B. A. Stewart (ed.), *Advances in Soil Science,* vol. 1. Springer-Verlag, NY.

Linell, K. A., and C. W. Kaplar. 1959. The factor of soil and material type in frost action. Bull. 255, Highway Res. Bd., Washington, DC, pp. 81–125.

Logsdon, S. D., J. C. Parker, and R. B. Reneau. 1987. Root growth as influenced by aggregate size. *Plant Soil* 99:267–275.

Low, A. J. 1955. Improvements in the structural state of soil under leys. *J. Soil Sci.* 6:177–199.

Low, A. J. 1972. The effect of cultivation on the structure and other physical properties of grassland and arable soils (1945–1970). *J. Soil Sci.* 23:363–380.

Low, P. F. 1980. The swelling of clay: II Montmorillonites. *Soil Sci. Soc. Amer. J.* 44:667–676.

Lynch, J. M. 1984. Interactions between biological processes, cultivation and soil structure. *Plant Soil* 76:307–318.

Lynch, J. M., and L. M. Panting. 1980. Cultivation and the soil biomas. *Soil Biol Biochem.* 12:29–33.

Mann, L. K. 1986. Changes in soil carbon storage after cultivation. *Soil Sci.* 142:279–288.

McGaw, R. 1972. Frost heaving versus depth to water table. *Highw. Res. Rec.* 393:45–55.

McGill, W. B., K. R. Cannon, J. R. Robertson, and F. D. Cook. 1986. Dynamics of soil microbial biomass and water soluble organic C in Breton L after 50 years of cropping to two rotations. *Canad. J. Soil Sci.* 66:1–19.

McKenzie, B. M., and A. R. Dexter. 1985. Mellowing and anisotropy induced by wetting of moulded soil samples. *Aust. J. Soil Res.* 23:37–47.

McKenzie, B. M., and A. R. Dexter. 1988. Radial pressures generated by the earthworm *Aporrectodea rosea*. *Biol. Fertil. Soils* 5:328–332.

Merckx, R., A. den Hortog, and J. A. van Veen. 1985. Turnover of root-derived material and related microbial biomass formation in soils of different texture. *Soil Biol. Biochem.* 17:565–569.

Mermut, A. R., and R. J. St. Arnaud. 1981. Microband fabric in seasonally frozen soils. *Soil Sci. Soc. Amer. J.* 45:578–586.

Metting, B. 1986. Population dynamics of *Chlamydomonas sajao* and its influence on soil aggregate stabilization in the field. *Appl. Envir. Microbiol.* 51:1161–1164.

Miller, R. D. 1980. Freezing phenomena in soils. pp. 254–299. In: D. Hillel (ed.), *Applications of soil physics*. Academic Press, NY.

Miller, W. P., and M. K. Baharuddin. 1986. Relationship of soil dispersibility to infiltration and erosion of southeastern soils. *Soil Sci.* 142:235–240.

Mirreh, H. F., and J. W. Ketcheson. 1973. Influence of soil water matric potential and resistance to penetration on corn root elongation. *Canad. J. Soil Sci.* 53:383–388.

Misra, R. K., A. R. Dexter, and A. M. Alston. 1986. Maximum axial and radial growth pressures of plant roots. *Plant Soil* 95:315–326.

Misra, R. K., A. M. Alston, and A. R. Dexter. 1988. Root growth and phosphorus uptake in relation to the size and strength of soil aggregates. I: Experimental studies. *Soil Tillage Res.* 11:103–116.

Mitchell, J. K. 1960. Fundamental aspects of thixotropy in soils. *J. Soil Mech. Found. Div.*, Amer. Soc. Civil Eng. 86(SM3):19–52.

Mitchell, J. K. 1976. *Fundamentals of soil behaviour*. Wiley, NY.

Mokhtaruddin, M. de Boodt. 1983. Evaluation of sensitivity to slaking by bulk density measurements of disrupted fragments (aggregates). *Pertanika* 6:90–94.

Molope, M. B. 1987. Soil aggregate stability: the contribution of biological and physical processes. *S. Afr. J. Plant Soil* 43:121–126.

Molope, M. B., and E. R. Page. 1986. The contribution of fungi, bacteria and physical processes in the development of aggregate stability of a cultivated soil. *Biol. Agr. Hort.* 3:233–249.

Molope, M. B., I. C. Grieve, and E. R. Page. 1985. Thixotropic changes in the stability of molded soil aggregates. *Soil Sci. Soc. Amer. J.* 49:979–983.

Morgenstern, N. R., and J. F. Nixon. 1971. One-dimensional consolidation of thawing soils. *Canad. Geotech. J.* 8:558–565.

Muneer, M., and J. M. Oades. 1989a. The role of Ca-organic interactions in soil aggregate stability. 1: Laboratory studies with ^{14}C-glucose, $CaCO_3$ and $CaSO_4 \cdot 2H_2O$. *Aust. J. Soil Res.* (in press).

Muneer, M., and J. M. Oades. 1989b. The role of Ca-organic interactions in soil aggregate stability. 2: Field studies with ^{14}C-labelled straw, $CaCO_3$ and $CaSO_4 \cdot 2H_2O$. *Aust. J. Soil Res.* (in press).

Muneer, M., and J. M. Oades. 1989c. The role of Ca-organic interactions in soil aggregate stability. 3: Mechanisms and models. *Aust. J. Soil Res.* (in press).

Nagasawa, T., and Y. Umeda. 1985. Effects of the freeze-thaw process on soil structure. In: *Proc. 4th Int. Symp. Ground Freezing,* Sapporo, Japan, vol. II, pp. 219–230.

Nikiforoff, C. C. 1941. Morphological classification of soil structure. *Soil Sci.* 52:193–212.

Oades, J. M. 1984. Soil organic matter and structural stability: mechanism and implications for management. *Plant Soil* 76:319–337.

Oades, J. M., A. G. Waters, A. M. Vassallo, M. A. Wilson, and G. P. Jones. 1988. Influence of management on the composition of organic matter in a red-brown earth as shown by ^{13}C nuclear magnetic resonance. *Aust. J. Soil Res.* 26:289–299.

Ojeniyi, S. O., and A. R. Dexter. 1979a. Soil factors affecting the macrostructures produced by tillage. *Trans. Amer. Soc. Agric. Eng.* 22:339–343.

Ojeniyi, S. O., and A. R. Dexter. 1979b. Soil structural changes during multiple pass tillage. *Trans. Amer. Soc. Agric. Eng.* 22:1068–1072.

Page, E. R. 1982. Soil conditioners. *Sci. Hort.* 33:1–13.

Pagliai, M., G. Guidi, M. La Marca, M. Giachetti, and G. Lucanante. 1981. Effects of sewage sludges and compost on soil porosity and aggregation. *J. Envir. Qual.* 10:556–561.

Panabokke, C. R., and J. P. Quirk. 1957. Effect of initial water content on the stability of soil aggregates in water. *Soil Sci.* 83:185–189.

Parton, W. J., D. S. Schimel, C. V. Cole, and D. S. Ojima. 1987. Analysis of factors controlling soil organic matter levels in Great Plains grasslands. *Soil Sci. Soc. Amer. J.* 51:1173–1179.

Pawluk, S. 1988. Freeze–thaw effects on granular structure reorganization for soil materials of varying texture and moisture content. *Canad. J. Soil Sci.* 68:485–494.

Perfect, E., B. D. Kay, W. K. P. van Loon, R. W. Sheard, and T. Pojasok. 1989a. Seasonal variation in soil structural stability under forages compared to corn. *Soil Sci. Soc. Amer. J.* (in press).

Perfect, E., B. D. Kay, W. K. P. van Loon, R. W. Sheard, and T. Pojasok. 1989b. Rates of change in soil structural stability under forages compared to corn. *Soil Sci. Soc. Amer. J.* (in press).

Power, J. F. 1987. *The role of legumes in conservation tillage systems.* Soil Cons. Soc. Amer., Ankenny, IA.

Powlson, D. S. 1980. The effects of grinding on microbial and non-microbial organic matter in soil. *J. Soil Sci.* 31:77–85.

Quirk, J. P., and B. G. Williams. 1974. The disposition of organic materials in relation to stable aggregation. *Trans. 10th Cong.,* Int. Soc. Soil Sci., Moscow, vol. 1, pp. 165–173.

Ramsay, A. J., R. E. Stannard, and G. J. Churchman. 1986. Effect of conversion from ryegrass pasture to wheat cropping on aggregation and bacterial populations in a silt loam soil in New Zealand. *Aust. J. Soil Res.* 24:253–264.

Reid, J. B., and M. J. Goss. 1982. Interactions between soil drying due to plant water use and decreases in aggregate stability caused by maize roots. *J. Soil Sci.* 33:47–53.

Reid, J. B., M. J. Goss, and P. D. Robertson. 1982. Relationship between the decrease in soil stability affected by the growth of maize roots and changes in organically bound iron and aluminium. *J. Soil Sci.* 33:397–410.

Rengasamy, P., R. S. B. Greene, G. W. Ford, and A. H. Mehanni. 1984. Identification of dispersive behaviour and management of red-brown earths. *Aust. J. Soil Res.* 22 413–431.

Rengasamy, P., R. S. B. Greene, and G. W. Ford. 1986. Influences of magnesium on aggregate stability in sodic red-brown earths. *Aust. J. Soil Res.* 24:229–237.

Rovira, A. D., and E. L. Greacen. 1957. The effect of aggregate disruption on the activity of microorganisms in soil. *Aust. J. Agric. Res.* 8:659–673.

Schubler, Prof. 1840. On the physical properties of soil, and on the means of investigating them. *J. R. Agric. Soc.* 1:177–218.

Scott. T. W., J. Mt. Pleasant, R. F. Burt, and D. J. Otis. 1987. Contributions of ground cover, dry matter, and nitrogen from intercrops and cover crops in a corn polyculture system. *Agron. J.* 79:792–798.

Scott Russell, R. 1977. *Plant root systems.* McGraw-Hill, Maidenhead, UK.

Shanmuganathan, R. T., and J. M. Oades. 1982. Effect of dispersible clay on the physical properties of the B horizon of a red-brown earth. *Aust. J. Soil Res.* 20:315–324.

Shanmuganathan, R. T., and J. M. Oades. 1983a. Influence of anions on dispersion and physical properties of the A horizon of a red-brown earth. *Geoderma* 29:257–277.

Shanmuganathan, R. T., and J. M. Oades. 1983b. Modification of soil physical properties by addition of calcium compounds. *Aust. J. Soil Res.* 21:285–300.

Shaviv, A., I. Ravina, and D. Zaslavsky. 1987. Application of soil conditioner solutions to soil columns to increase water stability of aggregates. *Soil Sci. Soc. Amer. J.* 51:431–436.

Shiel, R. S., M. A. Adey, and M. Lodder. 1988. The effect of successive wet/dry cycles on aggregate size distribution in a clay texture soil. *J. Soil Sci.* 39:71–80.

Shipitalo, M. J., and R. Protz, 1987. Comparison of morphology and porosity of a soil under conventional and zero tillage. *Canad. J. Soil Sci.* 67:445–456.

Shipitalo, M. J., and R. Protz. 1988. Factors influencing the dispersibility of clay in worm casts. *Soil Sci. Soc. Amer. J.* 52:764–769.

Shipitalo, M. J., R. Protz, and A. D. Tomlin. 1988. Effect of diet on the feeding and casting activity of *Lumbricus Terrestris* and *L. Rubellus* in laboratory culture. *Soil Biol. Biochem.* 20:233–237.

Sillanpaa, M., and L. R. Webber. 1961. The effect of freeze–thawing and wetting–drying cycles on soil aggregation. *Canad. J. Soil Sci.* 41:182–187.

Skidmore, E. L., and D. H. Powers. 1982. Dry soil aggregate stability: energy based index. *Soil Sci. Soc. Amer. J.* 46:1274–1279.

Skjemstad, J. O., and R. C. Dalal. 1987. Spectroscopic and chemical differences in organic matter of two vertisols subjected to long periods of cultivation. *Aust. J. Soil Res.* 25:323–335.

Skjemstad, J. O., R. C. Dalal, and P. F. Baron. 1986. Spectroscopic investigations of cultivation effects on organic matter of vertisols. *Soil Sci. Soc. Amer. J.* 50:354–359.

Soane, B. D. (ed.) 1982. *Compaction by agricultural vehicles: a review.* Scottish Inst. of Agric. Eng., Penicuik, Scotland.

Spoor, G. 1988. Improving the effectiveness of tillage operations. In: *Proc. Soil Mgt. '88,* Darling Downs Inst. Adv. Educ., Toowoomba, Queensland, Australia.

Spoor, G., S. M. Miller, and H. T. Breay. 1988. Timeliness and machine performance benefits from controlled traffic systems in sugar beet. pp. 317–322. In: *Proc. 11th Conf. Int. Soil Tillage Res. Org.,* Edinburgh, Scotland, vol. 1.

SSSA. 1975. *Soil Conditioners*. Soil Sci. Soc. Amer., Madison, WI.

Stefanson, R. C. 1968. Factors determining the seasonal changes in the stabilities of soil aggregates. *Trans. 9th Cong. Int. Soil Sci. Soc., Adelaide*, vol. 2. Angus and Robertson, Sydney, Australia, pp. 395–402.

Stengel, P. 1988. Cracks formation during swelling: effects on soil structure regeneration after compaction. pp. 147–152. In: *Proc. 11th Conf. Int. Soil Tillage Res. Org.*, Edinburgh, Scotland, vol. 1.

Stone, J. A., and B. R. Buttery. 1989. Nine forages and the aggregation of a clay loam soil. *Can. J. Soil Sci.* 69:165–169.

Stone, J. A., N. H. E. Allen, and C. D. Grant. 1985. Corn fertility treatments and the surface structure of a poorly drained soil. *Soil Sci. Soc. Amer. J.* 49:1001–1004.

Stypa, M., A. Nunez-Barrios, D. A. Berry, M. H. Miller, and W. A. Mitchell. 1987. Effects of sub-soil bulk density, nutrient availability and soil moisture on corn root growth in the field. *Canad. J. Soil Sci.* 67:293–308.

Taylor, J. H. 1987. A rationale for controlled traffic research. *Acta Hort.* 210:9–18.

Taylor, J. H., and E. C. Burt. 1987. Total axle load effects on soil compaction. *J. Terramechanics* 24:179–186.

Terpstra. R. 199. Formation of new aggregates under laboratory-simulated field conditions. Soil Tillage Res. 13:13–21.

Thomasson, A. J. 1978. Towards an objective classification of soil structure. *J. Soil Sci.* 29:38–46.

Tiessen, H., J. W. B. Stewart, and H. Hunt. 1984. Concepts of soil organic matter transformations in relation to organo-mineral plarticle size fractions. *Plant Soil* 76:287–295.

Tisdall, J. M., and H. H. Adem. 1986. Effect of water content of soil at tillage on size-distribution of aggregates and infiltration. *Aust. J. Exp. Agric.* 26:193–195.

Tisdall, J. M., and J. M. Oades. 1980a. The management of ryegrass to stabilize aggregates of a red-brown earth. *Aust. J. Soil Res.* 18:415–422.

Tisdall, J. M., and J. M. Oades. 1980b. The effect of crop rotation on aggregation in a red-brown earth. *Aust. J. Soil Res.* 18:423–433.

Tisdall, J. M., and J. M. Oades. 1982. Organic matter and water stable aggregates in soils. *J. Soil Sci.* 33:141–163.

Tisdall, J. M., B. Cockroft, and N. C. Uren. 1978. The stability of soil aggregates as affected by organic materials, microbial activity and physical disruption. *Aust. J. Soil Res.* 16:9–17.

Toogood, J. A., and D. L. Lynch. 1959. Effect of cropping systems and fertilizers on mean weight-diameter of aggregates of Breton plot soils. *Canad. J. Soil Sci.* 39:151–156.

Towner, G. D. 1986. The influence of particle size distribution on soil physical properties. *J. Agric. Sci. Camb.* 106:527–535.

Towner, G. D. 1987. The tensile stress generated in clay during drying. *J. Agric. Eng. Res.* 37:279–289.

Towner, G. D. 1988. The influence of sand and silt-sized particles on the cracking during drying of small clay-dominated aggregates. *J. Soil Sci.* 39:347–356.

Tu, J. D., and S. C. Tan. 1988. Soil compaction effect on photosynthesis, root rot severity and growth of white beans. *Canad. J. Soil Sci.* 68:455–457.

Tullberg, J. N. 1988. Controlled traffic in sub-tropical grain production. pp. 323–327. In: *Proc. 11th Conf. Intl. Soil Tillage Res. Org.*, Edinburgh, Scotland, vol. 1.

Utomo, W. H., and A. R. Dexter. 1981a. Tilth mellowing. *J. Soil Sci.* 32:187–201.

Utomo, W. H., and A. R. Dexter. 1981b. Age hardening of agricultural top soils. *J. Soil Sci.* 32:335–350.

Utomo, W. H., and A. R. Dexter. 1981c. Soil friability. *J. Soil Sci.* 32:203–213.

Utomo, W. H., and A. R. Dexter. 1982. Changes in soil aggregate water stability induced by wetting and drying cycles in non-saturated soil. *J. Soil Sci.* 33:623–637.

van Bavel, C. H. M., and F. W. Schaller. 1950. Soil aggregation, organic matter and yields in a long-term experiment as affected by crop management. *Soil Sci. Soc. Amer. Proc.* 15:399–404.

van der Linden, A. M. A., J. A. van Veen, and M. J. Frissel. 1987. Modelling soil organic matter levels after long-term applications of crop residues, and farmyard and green manures. *Plant Soil* 101:21–28.

van Ouwerkerk, C. 1968. Two model experiments on the durability of subsoil compaction. *Neth. J. Agric. Sci.* 16:204–210.

Voorhees, W. B. 1983. Relative effectiveness of tillage and natural forces in alleviating wheel-induced soil compaction. *Soil Sci. Soc. Amer. J.* 47:129–133.

Voorhees, W. B., and M. J. Lindstrom. 1984. Long-term effects of tillage method on soil tilth independent of wheel traffic compaction. *Soil Sci. Soc. Amer. J.* 48:152–156.

Voorhees, W. B., C. G. Senst, and W. W. Nelson. 1978. Compaction and soil structure modification by wheel traffic in the northern corn belt. *Soil Sci. Soc. Amer. J.* 42:344–349.

Warren, S. D. 1987. Soil hydrologic response to intensive rotation grazing: a state of knowledge. pp. 488–501. In: Y. S. Fok (ed.), *Proc. Int. Conf. on Infiltration Development and Application*. Water Resources Res. Center, Univ. of Hawaii.

Warren, S. D., T. L. Thurow, W. H. Blackburn, and N. E. Garza. 1986. The influence of livestock trampling under intensive rotation grazing on soil hydrologic characteristics. *J. Range Mgt.* 39:491–495.

Willis, W. 1955. Freezing and thawing, and wetting and drying in soils treated with organic chemicals. *Soil Sci. Soc. Amer. Proc.* 19:263–267.

Yamada, S. 1955. Investigation on freezing and thawing actions of volcanogenous soils and their influences upon soils and crops, pt. 1. *Soil and Plant Food* 1:84–87.

Yeoh, N. S., and J. M. Oades. 1981. Properties of clays and soils after acid treatment, II: Urrbrae fine sandy loam. *Aust. J. Soil Res.* 19:159–166.

Zachman, J. E., D. R. Linden, and C. E. Clapp. 1987. Macroporous infiltration and redistribution as affected by earthworms, tillage and residue. *Soil Sci. Soc. Amer. J.* 51:1580–1586.

Crop Responses to *Azospirillum* Inoculation

M.E. Sumner*

I. Introduction

Following the first reports of an *Azospirillum*–plant association over the last decade, there has been considerable interest in the use of *Azospirillum* species as nitrogen-fixing bacteria in association with the roots of monocotyledonous plants, particularly grasses and grain crops. The original thesis was that *Azospirillum* (synonym *Spirillum*) invades cortical cells, forming a "primitive symbiotic association" which was suspected of fixing N_2 (Dobereiner and Day, 1976). Despite a substantial body of literature, there is still a lack of clarity concerning the mechanisms that result in improved growth and performance of the host plant, or why lack of response is often observed in the field. This is not altogether surprising because this phenomenon is complex and quite variable.

*Department of Agronomy, University of Georgia, Athens, GA 30602, U.S.A.

There is a definite need for the identification of the most critical parameters for beneficial association.

In the discussion to follow, the successes and failures of *Azospirillum* inoculation in relation to N_2 fixation and hormonal and other effects will be analyzed. The literature will be reviewed from the agronomic perspective so that a conclusion may be reached as to whether or not *Azospirillum* inoculation will become a feasible and profitable technique in the near future.

Before proceeding to an evaluation of greenhouse and field growth responses, it is necessary to review briefly the interaction of *Azospirillum* with roots. In this manner, a better understanding of growth responses and the resultant morphological and chemical changes may be obtained.

II. Interactions of *Azospirillum* with Roots

A. *Azospirillum* spp.

Azospirillum is ubiquitous in the rhizosphere of grasses in the field. Usually two species are found, *A. lipoferum,* most commonly as a surface-sterile isolate from the roots of C-4 grasses, and *A. brasilense* from C-3 grasses and sugarcane (C-4) in tropical and subtropical regions. However, Horemans et al. (1988) have recently found no evidence for the presence of *Azospirillum* within the roots of barley, wheat, corn, or grasses grown under field conditions in Belgium, with *A. lipoferum* being found only on the root surfaces of corn and grass and *A. brasilense* on the roots of all crops investigated. Concentrations of these bacteria never exceeded 10^3 colony-forming units (cfu) per gram of soil. *Azospirillum* is microaerophilic and shows a marked preference for low p_{O_2} with very little nitrogenase activity at p_{O_2} values above 1 kPa (Patriquin et al., 1983). The main difference between the two species is the ability of *A. lipoferum* to produce acid in the fermentation of sugars, whereas *A. brasilense* possesses primarily an oxidative metabolism. Under microaerophilic conditions *Azospirillum* fixes N_2, but under aerobic conditions it is capable of utilizing combined inorganic and organic forms of N. Several strains are capable of denitrification, which is favored under the same conditions as those that promote nitrogenase activity (Danneberg et al., 1985; Hubbell and Gaskins, 1984).

B. Adsorption to Plant Roots

Because plant roots excrete organic and inorganic substances suitable as nutritional sources for microorganisms, it is not surprising to find elevated numbers of bacteria in the rhizosphere of plants. On the one hand, in highly specific systems, such as the *Rhizobium*–legume association, large numbers of bacteria are selectively adsorbed on the root hairs of the homologous host, where they can be weakly or strongly attached, with the latter competing better in colonization and nodulation (Vance, 1983). On the other hand, in experiments with nonspecific bacteria adsorption did not follow the Langmuir adsorption isotherm, whereas with *Rhizobium,* proper isotherms were obtained and the adsorption appeared to be via a specific protein (Shimshick and Herbert, 1979).

In the case of *A. brasilense* strain Sp 7, polar adsorption to root hairs and old epidermal cells was rapid, with age of culture and suspending electrolyte affecting the number of bacteria adsorbed. After several hours, fibrillar structures appeared as anchors to the root. Azospirilla were firmly attached to root hairs of grasses grown in the absence but not in the presence of fixed nitrogen, in which case they were associated only with the mucigel at the root cap and with undifferentiated epidermal cells. Significantly fewer azospirilla attached to millet roots when grown in media containing 5 m*M* potassium nitrate (Umali-Garcia et al., 1980). Several workers (Kapulnik et al., 1985a; Umali-Garcia et al., 1980, 1981) showed that the adsorption of different strains of *A. brasilense* to wheat *(Triticum aestivum)*, pearl millet *(Pennisetum americanum)*, and guinea grass *(Panicum maximum)* was greatly affected by the growth phase of the culture. The attachment properties of different *Azospirillum* strains on wheat roots suggests that they are affected at the bacterial genome level (Jain and Patriquin, 1984). Presumably, mechanisms by which *Azospirillum* attaches to root hairs and undifferentiated epidermal cells differ in response to changes caused by fixed nitrogen in the rooting medium (Patriquin et al., 1983; Umali-Garcia et al., 1980, 1981). More recently, adsorption of *A. brasilense* strain Cd on corn *(Zea mays)* roots increased rapidly for 90 minutes, reaching a maximum after 4.5 hours at a pH of 6.1, with other strains being adsorbed to different extents, each having its own optimum pH value (Okon and Kapulnik, 1986). Bacterial adsorption occurred on cells in the root elongation zone and at the bases of root hairs and followed the Langmuir isotherm up to 10^9 cells/mL, above which larger numbers than expected were adsorbed, probably due to the adsorption of aggregates of cells. Binding of *A. brasilense* strain Cd was reduced in the presence of Ca, Mg, Na, poly (L-lysine), several sugars, and amino and organic acids at pH 6.1 with evidence of bacterial response to chemotactic gradients. This possibly provides the bacteria with the ability to reach and preferentially colonize plant roots. Hess et al. (1985) demonstrated that sucrose, glucose, and fructose attracted azospirilla to wheat roots with the chemotactic potential being increased by sucrose cleavage by an invertase secreted by the roots. Azospirilla adhered to pearl millet roots in greater numbers than all other bacteria examined. This indicates that grass root hairs exhibit selectivity in binding bacteria (Table 1). *R. trifolii* and *P. fluorescens* also adhere to millet roots and, interestingly, they share certain serological and biochemical characteristics with *Azospirillum*. Adsorption of *R. japonicum* to wheat and rice *(Oryza sativa)* roots has been reported previously (Shimshick and Herbert, 1979). Preincubation of *A. brasilense* with root exudates promoted adsorption. This effect was negated by treatment of the extracts with enzymes (protease and trypsin), which suggests the possible involvement of proteins-lectins in the adsorption–recognition process of *Azospirillum* to grass roots. Indeed, *A. lipoferum* strains 4B and A-95 stimulated root exudation in rice plants, whereas strains RO7 and B7C did not (Heulin et al., 1987). No information is available on the effect of random or specific adsorption on colonization and proliferation on and in the roots (Okon and Kapulnik, 1986), but Eyers et al. (1988) have shown that live *A. brasilense* strains Sp 7 and Sp 245 attach to plant cells more strongly than to dead cells.

Table 1. Adsorption of various bacteria to root hairs of pearl millet *(Pennisetum americanum)* seedlings

Bacterium root	Number of root hairs examined	Number of adsorbed cells per hair
Azospirillum brasilense strain Sp 13t	16	23.7 ± 2.4
Azospirillum brasilense strain Sp 7	18	33.0 ± 4.0
Azospirillum brasilense strain JM 125A2	19	32.5 ± 2.9
Rhizobium trifolii strain 0403	19	14.4 ± 2.0
Pseudomonas fluorescens	15	11.4 ± 2.9
Azotobater vinelandii strain UW10	15	0.6 ± 0.3
Klebsiella pneumoniae	18	0.3 ± 0.2
Escherichia coli	16	0.1 ± 0.1

From Umali-Garcia et al. (1980).

C. Root Colonization

It is well known that in the *Rhizobium*–legume association and with root pathogens there is considerable specificity in the infection of the host, which is also likely to be the case in *Azospirillum*–gramineae associations.

In order to obtain yield responses to field inoculation with *Azospirillum*, it is essential that the strain or strains used establish themselves and multiply in the rhizosphere or in roots where indigenous strains and other organisms offer stiff competition. The enrichment of the rhizosphere in *Azospirillum* relative to the bulk soil can be brought about by any one or a combination of the following: (1) competitive advantage in the carbon-rich, nitrogen-poor environment, (2) supply by the plant of essential vitamins, (3) aerotactic attraction to reduced p_{O_2} in the root zone (Patriquin et al., 1983), and (4) chemotactic attraction to root exudates and organic acids (Baldani et al., 1986). In general, while bacteria are randomly distributed on grass roots, *Azospirillum* colonization usually occurs on mucigel-covered epidermis, in the elongation zone, on the bases of root hairs, and where root branches emerge. Bacteria were found in the mucigel that accumulated on the root caps and along the root axes, but fewer bacteria were present on the root cap or adsorbed to root hairs of wheat (Kapulnik et al., 1985a). The viscous nature of mucigel would provide a barrier to gas diffusion, which may create a microaerophilic environment conducive to N_2 fixation. Adherent bacteria were associated with granular material on root hairs and fibrillar material on undifferentiated epidermal cells (Okon and Kapulnik, 1986; Umali-Garcia et al., 1980). Tilak and Subba Rao (1987) found that *Azospirillum* occurred as encrustations in the mucilage, colonizing root hairs and the innermost layer of cortex, and conducting vessels in pearl millet with almost total absence toward the root tip. Relatively large populations of *Azospirillum*, contributing significantly to microbial biomass, are now known to invade root interiors, where they reside between cortical cells and in apparently dead cells (Baldani and Dobereiner, 1980; Patriquin et al., 1983; Umali-Garcia et al., 1980), although Reinhold et al.,

(1986) found evidence of close association between *A. lipoferum* and a new type of *Azospirillum* spp. and kallar grass *(Leptochloa fusca)* on the rhizoplane with unidentified diazotrophic straight rods in the endorhizosphere. This distribution of the populations was stable, suggesting that different diazotrophs may be adapted to colonize different root zones of their hosts. Bashan and Levanony (1988) showed that after inoculation of wheat with *A. brasilense* strain Cd, small aggregates of bacteria formed on the root surface, with a larger population within the cortex and no bacteria in the xylem. A few bacteria penetrated young live root cells, but the majority occurred in intercellular spaces or adsorbed to cortical or epidermal cell walls by an electron-dense material. As a result of the inoculation, proton exudation and cell division at the root tip were increased. As *Azospirillum* has pectinolytic activity (see later), it is not surprising that it can penetrate root hairs and other cells. However, no evidence of the formation of viable, intracellular associations with grasses has been found. The importance of these azospirilla within the root is indicated by the fact that in wheat, total N accumulation was correlated with numbers recovered from surface-sterilized but not untreated roots (Baldani et al., 1983). In rice inoculated with *A. lipoferum* strain 34-H, the root tips of one cultivar (IR42) were not initially colonized, whereas those of another (IR50) were. In the early stages, most of the bacteria were embedded in the ruptured mucigel below the root cap cells of IR42, while mucigel was hardly detectable in IR50 (Murty and Ladha, 1987). The root hair primordia of IR50 were more rapidly colonized than those of IR42. Thus adherence of bacteria may depend on the nature of the root surface rather than on the presence or absence of mucigel. It was suggested that *Azospirillum* invades the intercellular spaces in the inner cortex and stele and the lumen of the xylem vessels without disrupting the outermost cortex or endodermis. It appears that the initial entry into the stele occurs at apical regions where the endodermal walls and Casparian strip have not thickened and where the stele has not differentiated. This assumes that the normal processes which resist bacterial invasion are not operative at the root tip, which was the site of initial colonization of soybean *(Glycine max)* roots. This internal association between *Azospirillum* and the host would facilitate an efficient exchange of materials, and the presence of substrate and low p_{O_2} in the xylem would be conducive to N_2 fixation, but the low pH of the xylem sap would tend to prevent this process (Patriquin et al., 1983).

While there are no observations suggesting that N_2 fixation or yield is related to any specific sites of *Azospirillum* colonization, substantial evidence exists to suggest some internal localization or protective structure:

1. Microbial populations obtained from surface-sterilized roots are different from untreated roots, with the former being enriched in azospirilla. No consistent predominance of either *Azospirillum* species occurred when the bacteria were isolated from washed roots of rice, wheat, or maize. However, after surface sterilization with chloramine-*t*, *A. lipoferum* was dominant in isolates from maize and other C-4 graminaceous crops (except sugarcane [*Saccharum officinarum*]) and *A. brasilense* nir$^-$ from rice, wheat, and other C-3 crops (Baldani and dobereiner, 1980).

2. Intact cortex and increased root activity with increasing lateral roots are consistent with colonization through root tips (Dobereiner and Day, 1976).
3. High apparent rate constants for acetylene reduction activity (ARA) and postlag ARA insensitivity to p_{O_2} are consistent with the presence of barriers to gaseous diffusion to and from N_2-fixing sites (Patriquin, 1978).
4. Total numbers of *Azospirillum* within the root are correlated with total N accumulation, which is not the case with non-surface-sterilized roots (Baldani et al., 1983).
5. N_2 fixation is low in leaky and high in nonleaky roots of wheat lines (Rennie and Larson, 1979).
6. "C forms," spheres, and other structures formed by *Azospirillum* are bound by a membrane to the root surfaces of wheat (Patriquin et al., 1983). This would afford some degree of protection from external oxygen.
7. The effect of metabolic poisons applied to the root surface is consistent with the internal localization of some N_2-fixing sites (Boyle and Patriquin, 1980).
8. Direct evidence from electron microscopy shows internal colonization by a limited number of bacteria, with a zone relatively free of bacteria separating the majority of the bacteria from the root surface (Whallon et al., 1985).

Reinhold et al. (1986) suggested that, since large numbers of N_2-fixing bacteria occur within the root, which may result in an advantage for substrate competition against nondiazotrophs, endorhizosphere bacteria might be more important for N supply to the host than rhizoplane bacteria. However, no quantitative estimates of *Azospirillum* biomass on or in the roots are available (Okon et al., 1988).

The effect of encapsulation on the attachment process is not well understood, but *Azospirillum* forms various structures like *Rhizobium,* which are thought to confer oxygen protection and resistance to desiccation (Patriquin et al., 1983). However, none of these structures are visible to the eye, and thus successful infection cannot be evaluated easily. When incubated for prolonged periods, *Azospirillum* produces cystlike ovoids, which may provide a mechanism for survival under adverse conditions in the rhizosphere (Hubbell and Gadkins, 1984). Sadasivan and Neyra (1985) demonstrated the formation of desiccation-resistant cysts in *A. brasilense* strain Sp 7 and *A. lipoferum* strain Sp 59b associated with brown-black melaninlike granular pigments under nitrogen-limiting conditions. The formation of these cysts appears to be controlled by genetic and environmental factors.

Thus plant roots respond in different ways to varying levels and kinds of azospirilla, but, in general, the effect seems similar to the symptoms obtained when plants are infected with varying levels of pathogenic bacteria. There is evidence that a recognition process is involved in colonization similar to the lectin-recongition model in the *Rhizobium*–legume association (Umali-Garcia et al., 1980). Colonization probably takes place at specific sites, favored by an optimal inoculum level in each case. In addition, there is probably substantial competition for sites by other rhizosphere organisms. Furthermore, evidence indicates that nitrate and, to a lesser extent, ammonium strongly inhibit attachment of nonhomologous but not homologous strains to maize roots.

Schmidt et al. (1988) showed that inoculation of alfalfa *(Medicago sativa)* with a combination of *Rhizobium meliloti* and *A. brasilense* strain Cd in sterile agar resulted in a significant increase in the number of nodules depending on the concentration of *Azospirillum*. The stimulating effect is probably due to the production of indol-3-acetic acid (IAA), which in pure form leads to increased nodulation. An increase in nodule number was also found with soybeans grown in soil under nonsterile conditions upon repeated inoculation with *Azospirillum*. *Azospirillum brasilense* was much more effective in producing IAA than *A. lipoferum*. Yahalom et al. (1987), working with pouch-grown *Medicago polymorpha* and *Macroptilium atropurpureum* seedlings, showed the preinoculation with *A. brasilense* strain Cd 24 hours before Rhizobium resulted in increased nodulation and ARA. The stimulation of a larger number of epidermal cells that differentiate into infectable roots by the *Azospirillum* inoculation was advanced as a possible reason for the increased susceptibility to *Rhizobium* infection. Similar results were obtained with various forage legumes grown in pot culture (Yahalom et al., 1988). On the other hand, Raverkar and Konde (1988) found that inoculation of peanuts *(Arachis hypogaea)* with *A. lipoferum* and *Rhizobium* strain NC92 singly increased nodule number, nodule dry weight, amount of dry matter produced, and N content, but inoculation with a mixture of these organisms largely negated the effect on pod yield and decreased it with respect to the other parameters.

D. Establishment and Proliferation in the Field

Relatively little information has been published on the establishment and multiplication of *Azospirillum*, especially under field conditions. Following inoculation, many strains survive in the rhizosphere, but invariably populations stabilize at low levels. Albrecht et al. (1983) and Smith et al. (1984b) reported that *A. brasilense* strains 13t SR2 and CdSR failed to establish in the rhizosphere of sorghum *(Sorghum bicolor)*, maize, and pearl millet in Florida. In extensive field experimentation in Brazil, Baldani et al. (1986) demonstrated that *A. brasilense* strains Sp 107 and Sp 245 could be established in all wheat experiments and were the dominant strains in washed and especially surface-sterilized roots, whereas strains Sp 7 and Cd established poorly. *Azospirillum lipoferum* strain Sp S82 was the dominant root isolate from sorghum inoculated with this strain, which became concentrated in the upper parts of the root system. On the other hand, Harris et al. (1989) showed that *Azospirillum* is not a good colonizer of wheat rhizospheres after inoculation under temperate conditions. Thus no physical or chemical treatments appear to bring about the preferential increase in *Azospirillum* numbers in soil. Whether this will ever become possible is not clear, but a more likely possibility would involve specificity between bacterium and host that would provide *Azospirillum* with a competitive edge. However, there still remain many technical problems to be overcome before proper colonization of roots can be assured under a variety of soil and environmental conditions. The use of pesticides may have an effect on *Azospirillum* performance, but to date no field studies have been carried out. Herbicides such as metrabuzin and ethiozin have short-term inhibitory effects on the nitrogenase activity of *A. brasilense* and *A. lipoferum* under laboratory conditions, but the effects under field conditions are unknown (Gadkari, 1988). The organochloride insecticide Difocol at concentrations less than 1 ppm under laboratory conditions has been shown to stimulate the nitrogenase activity of *A. lipoferum* strain Sp Br17, but above 10 ppm cell viability and nitrogenase activity were severely reduced, resulting in the formation of cystlike structures which may reflect an induced resistance mechanism (Mano et al., 1988).

E. *Azospirillum* and Root Morphology and Function

1. Root Hair Deformation

Inoculation with *Azospirillum* spp. has a profound effect on root morphology and function. Root hair numbers and differential root lengths are increased, with individual strains showing effects of differing intensity. Basically two types of branching, namely, "tuning fork deformation" with branches of equal length and branches of unequal length, were observed in field-grown wheat seedlings inoculated with *A. brasilense,* with strain Sp 245 being more effective than strains Sp 7 and Sp 107. High frequencies of deformed wheat and maize root hairs were observed in semisolid agar but not in water agar (Patriquin et al., 1983; Tien et al., 1979). In semisolid and water agar, *Azospirillum* seems to promote tuning fork deformations with the effect in *A. brasilense* strain Sp 245 > strain Sp 107 > strain Sp 7 > *Azospirillum* spp. strain Sp 242, which corresponds to the order of total N accumulation in wheat following inoculation of field-grown wheat (Baldani et al., 1983). Few or no tuning fork deformations are produced by nonhomologous strains Sp 7 and Sp 242. This suggests that tuning fork deformations may have some prognostic value. The number of deformations was higher in the presence than in the absence of N, which is contrary to the effect in legumes. Root hair deformations were also observed in field-grown maize, but deformations could not be induced in agar cultures, even with homologous strains of *Azospirillum*. If hydrolytic enzymes are responsible for root hair deformations in wheat, as the lectin-enzyme hypothesis of Hubbell (1981) would suggest, then root hair deformation may be related to the ability of strains to invade the root. This is supported by the observations of a high degree of activity present in populations isolated from surface-sterilized but not from untreated roots (Baldani and Dobereiner, 1980; Baldani et al., 1983; Patriquin et al., 1983). In a number of crop plants, *Azospirillum* inoculation at 10^5 cfu/mL resulted in denser and longer roots hairs (2.5 cm from the root cap) than in controls treated with dead cells. This did not affect the length of the elongation zone, which suggests that the effect is due to an earlier initiation of root hair formation. At 10^8 cfu/mL, the elongation zone length was greatly decreased with severe deformation of the root cap. In *Sorghum vulgare* × *Sorghum sudanense* inoculated with *A. brasilense* strain Cd, the distance between root cap and hairs was significantly greater than in untreated controls at 10^5 cfu/mL, but it was significantly smaller at 10^8 cfu/mL (Okon and Kapulnik, 1986).

In general, effects on root hair development are more marked at high inoculum concentrations, which might explain why root hair deformations were observed in semisolid agar, which would favor bacterial proliferation, but not in water agar (Patriquin et al., 1983). It is not clear whether the effects on root hair formation are caused by root cap growth inhibition effects on cell division or by elongation at the elongation zone, or whether increased root hair formation is caused by the differentiation of cells in the elongation zone to form root hairs. All these effects depend on and vary with inoculum concentration, bacterial and plant type, method of growing the plant, and environmental conditions (Okon and Kapulnik, 1986).

2. Root Elongation and Surface Area

Root elongation was increased in a number of crops (*Pennisetum, Setaria italica,* and wheat) by *Azospirillum* inoculation under both greenhouse and field conditions. With a single wheat cultivar at 10^5 to 10^6 *Azospirillum* cfu/mL, both root length (Figure 1) and total surface area reached a maximum (Kapulnik and Okon, 1983), and the magnitude of the effect varied according to culture age (lag phase cultures being less effective) and incubation temperature, but 10^8 to 10^{10} cfu/mL inhibited root development. With a mixture of *Azospirillum* strains and a variety of wheat cultivars, inoculation caused an increase in root length in all cultivars, but increased surface area only in some (Figure 2). Other bacteria in the genera *Klebsiella, Azotobacter, Bacillus,* and *Pseudomonas* did not increase the surface area of wheat roots (Kapulnik and Okon, 1983; Okon and Kapulnik, 1986; Tien

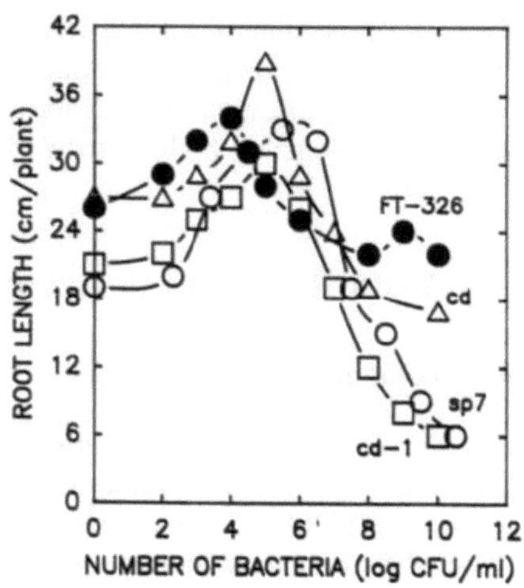

Figure 1. Effect of inoculum concentration of different strains of *Azospirillum* on root elongation of wheat seedlings cultivar Miriam. ● FT-326, IAA overproducing mutant; △ Cd; ○ Sp 7; □ Cd-1. [From Kapulnik et al. (1985a).]

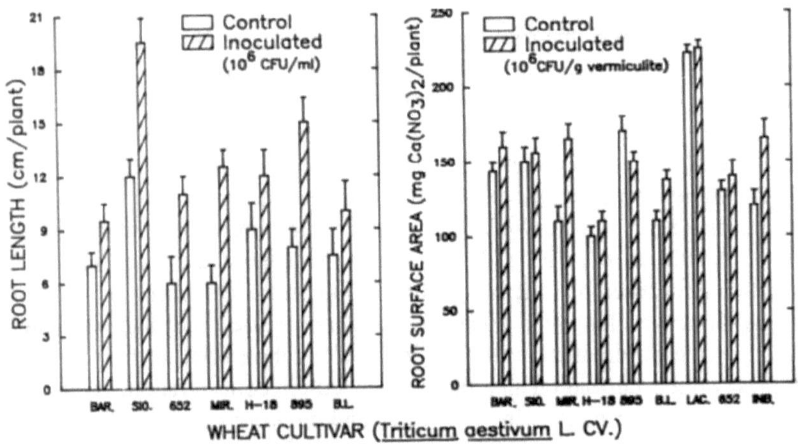

Figure 2. Effect of *Azospirillum*-mixed inoculum on root length and surface area of different wheat cultivars. [From Kapulnik et al. (1985a).]

et al., 1979; Yahalom et al., 1984) (Table 2). In C-3 (barley [*Hordeum vulgare*] and wheat) and C-4 *(Panicum miliaceum)* grasses, different and mixed strains of *Azospirillum* increased the elongation of seminal roots with bacterial concentration. Elongation was optimal in the range of 10^3 to 10^6 cfu/mL, above which roots were inhibited. Optimal bacterial concentrations and critical levels for inhibition varied for the different species and strains tested (Okon and Kapulnik, 1986). These data suggest that *Azospirillum* behaves in a manner similar to phytopathogenic bacteria because of the relationship between bacterial concentration and development of symptoms. As specific sites probably exist on roots where colonization is initiated, there is likely to be an optimal inoculum concentration for eliciting positive responses in root growth and development. Because the optimal inoculum level will depend on *Azospirillum* strain, plant type, and environmental conditions, these factors must be considered in developing inoculants and application methods for introducing the required number of appropriate root-colonizing bacteria for different types of soil.

III. Yield Responses to *Azospirillum* Inoculation

A. Greenhouse and Laboratory

A summary of the yield responses to *Azospirillum* of many crops published in the literature is presented in Table 3 with relevant information on N content, N uptake, and ARA. The responses fall into the following broad categories: (1) those where only yield increases without pertinent measures of N_2 fixation were

Table 2. Comparison of the effect of *Azospirillum* and other bacteria on wheat root development

Bacterium	Concentration (cfu/mL)	Increase in root length (%)	Increase in root surface area (%)
Klebsiella pneumoniae	10^6	6.8 ns[a]	18.2 ns
	10^8	6.8 ns	15.8 ns
Azotobater chroococcum	10^6	27.6 *[b]	–4.0 ns
	10^8	20.6 *	–4.0 ns
Azospirillum	10^6	20.6 *	33.7 *
	10^8	1.3 ns	2.1 ns
Bacillus megaterium	10^6	13.8 ns	–5.4 ns
	10^8	2.7 ns	–5.8 ns
Pseudomonas syringae	10^6	6.9 ns	6.7 ns
	10^8	–11.0 ns	10.3 ns
Control (dead *Azospirillum* cells)		—	—

From Kapulnik and Okon (1983).

[a]ns = No significant difference.

[b]* = Significant at the 0.05 probability level.

recorded [Jagnow (1985) with *Arrhenatherum elatius* and *Festuca rubra;* Kapulnik et al. (1981c, 1985c) with *Setaria italica* and *Triticum aestivum* [roots only]; Lin et al. (1983) with *Zea mays* and *Sorghum bicolor;* Millet et al. (1985) with *Triticum aestivum* and *Triticum turgidum;* Nayak et al. (1986) with *Oryza sativa;* Tien et al. (1979) with *Pennisetum americanum;* and Subba Rao et al. (1979) with *Oryza sativa]*, (2) those where no yield responses were obtained to inoculation [Albrecht et al. (1981) with *Zea mays;* Baltensperger et al. (1978) with *Cynodon dactylon;* Bouton and Zuberer (1979) with *Panicum maximum;* Haahtela et al. (1988) with *Poa pratensis* and *Triticum aestivum;* and Schank et al. (1981) with *Digitaria* spp.], and (3) those where positive yield responses and increased N uptake, or increased ARA, or both, have been observed [Avivi and Feldman (1982) with *Triticum durum;* Bouton and Zuberer, (1979) with *Panicum maximum;* Cohen et al. (1980) with *Setaria italica* and *Zea mays;* Fayez and Daw (1987) with *Gossypium barbadense;* Fayez et al. (1985) with wheat; Hadas and Okon (1988); Hegazi et al. (1983) with *Zea mays;* Kapulnik et al. (1981a, 1981b) with *Panicum miliaceum, Sorghum bicolor* × *Sorghum sudanense, Sorghum bicolor,* and *Setaria italica;* Lethbridge and Davidson (1983) with *Triticum aestivum;* Malik et al. (1987) with *Leptochloa fusca;* Mertens and Hess (1984) with *Triticum aestivum;* Nayak et al. (1986) with *Oryza sativa;* Nur et al. (1980) with *Zea mays* and *Setaria italica;* O'Hara et al. (1981) with *Zea mays;* Rai and Gaur (1988) with *Triticum aestivum;* Rennie (1980) with *Zea mays;* Rennie and Larson (1979) with *Triticum aestivum* (disomic chromosome substitution lines); Smith et al. (1976) with *Digitaria decumbens;* Venkateswarlu and Rao (1983) with *Pennisetum americanum;* Yahalom et al. (1984) with *Setaria italica;* and Watanabe and Lin (1984) with *Oryza sativa]*. Where N uptake or ARA was measured, no cases were observed where inoculation caused a decrease in either of these values. There are cases where specific strains are required to elicit a yield response [Cohen et al. (1980) with *Zea mays* and *Setaria italica;* Fayez and Daw (1987) with *Gossypium barbadense;* Ferreira et al. (1987) with *Triticum aestivum;* Lin et al. (1983) with *Sorghum bicolor;* O'Hara et al. (1981) with *Zea mays;* O'Hara et al. (1987) with *Zea mays* and *Pennisetum americanum;* Rai and Gaur (1988) with *Triticum aestivum;* and Venkateswarlu and Rao (1983) with *Pennisetum americanum]*, pointing to the importance of inoculating crops with homologous strains of *Azospirillum*. In certain cases, increased root growth accompanied the yield response, but this was not consistent in all cases. From the results presented in Table 3, it can be safely concluded that, provided an appropriate inoculating strain is used, *Azospirillum* inoculation results in increased yields and N uptake for a wide range of tropical and temperate crops under a wide range of controlled conditions in the greenhouse and the laboratory. There is also substantial evidence that inoculation with *Azospirillum* increases rooting, as will be discussed later. It is pertinent to note that relatively few cases have been reported where no yield responses were obtained to inoculation. This does not necessarily imply that such cases do not exist.

B. Field

It is encouraging to discover that many workers have conducted field inoculation experiments with *Azospirillum*. This is relatively rare for such a recently discovered phenomenon. Results are summarized in Table 4 with pertinent data on N content, uptake, and fixation. These results fall into the following categories: (1) those where inoculation failed to elicit a yield response [Baldani et al. (1987) with *Triticum aestivum;* Marocco and Lorenzoni (1984) with *Triticum aestivum;* Millet et al. (1985) with *Triticum aestivum;* Taylor (1979) with *Panicum maximum;* Tilak and Subba Rao (1987) with *Pennisetum americanum;* Wani et al. (1985) with *Pennisetum americanum;* and Zambre et al. (1984) with *Triticum durum]*, (2) those where yield responses were obtained but no accompanying N data were presented [Bouton et al. (1979) with *Pennisetum americanum;* Charyulu et al. (1985) with *Oryza sativa;* Hegazi and Saleh (1985) with *Triticum aestivum;* Kannaiyan et al. (1983) with *Oryza sativa;* Kapulnik et al. (1983) with *Triticum aestivum;* Kapulnik et al. (1987) with *Triticum aestivum;* Marocco and Lorenzoni (1984) with *Hordeum vulgare;* Okon et al. (1988b) with *Zea mays, Triticum aestivum, Sorghum bicolor, Setaria italica,* and *Panicum miliaceum;* Rao et al. (1983) with *Oryza sativa;* Smith et al. (1976) with *Pennisetum americanum* and *Panicum maximum;* Taylor (1979) with *Pennisetum americanum;* and Tilak and Subba Rao; (1987) with *Pennisetum americanum]*,(3) those where positive yield responses were presented with supporting data [Albrecht et al. (1981) with *Zea mays;* Boddey et al. (1986) with *Triticum*

Table 3. Greenhouse yield responses of various crops to inoculation with *Azospirillum* spp. together with corresponding N content and uptake and acetylene reduction activity

Crop	Medium	Bacterium	Inoculation	Fertilizer N	Yield	Roots	N fixation	N uptake	^{15}N uptake	N content (%)	Reference
			<------	------	------	------ various units[a] ------	------	------	------	------>	
Digitaria decumbens	Soil	*A. lipoferum* Sp 13t	No	0	180			1.90		1.06	Smith et al. (1976)
			Yes	0	300*			3.07		1.02	
Panicum maximum			No	0	800			11.01		1.38	
			Yes	0	1228*			20.05		1.63	
Cynodon dactylon	Soil	*A. brasil/paspali*	Dead	0	5.40	3.05		40.50			Baltensperger et al. (1978)
			Live	0	5.63	2.88		49.00*			
Pennisetum americanum	Solution	*A. brasilense* Sp 13t SR2	No	0	52.67	12.01					Tien et al. (1979)
			Yes	0	74.47*[b]	12.05					
			No	21	694	161					
			Yes	21	1017*	211					
Panicum maximum	Soil	*A. brasilense* Sp 13t	Water	60	2.68		71				Bouton and Zuberer (1979)
			Live	60	3.50*		100				
			Filtrate	60	3.43		90				
			Water	40	2.10		96				
			Live	40	1.81		152				
			Filtrate	40	1.70		45				
Oryza sativa	Soil	*A. brasilense*	No	0	15.8						Subba Rao et al. (1979)
			Yes	0	20.9*						
			No	40	33.9						
			Yes	40	41.3*						
			No	60	48.1						
			Yes	60	54.4*						
			No	120	68.9						
			Yes	120	70.6						

Table 3. *Continued*

Crop	Medium	Bacterium	Inocu-lation	Ferti-lizer N	Yield	Roots	N fix-ation	N uptake	^{15}N uptake	N con-tent (%)	Reference
			<------	------	------	------	various units[a]	------	------	------>	
Triticum											Rennie and
aestivum	Solution	*A. brasilense*	No	100	287	101		14.88			Larson (1979)
Cadet		Sp 7	Yes	100	436	172*		22.51*			
C-R2A			No	100	256	116		12.90			
			Yes	100	543*	185*		28.00*			
C-R2D			No	100	408	160		19.04			
			Yes	100	748*	136		32.60*			
C-R5B			No	100	389	77		14.77			
			Yes	100	413	98		19.03*			
C-R5D			No	100	320	124		15.11			
			Yes	100	405*	142		20.09*			
Rescue			No	100	372	134		20.14			
			Yes	100	349	156		18.11			
R-C2A			No	100	509	81		21.80			
			Yes	100	429*	146*		20.34			
R-C2D			No	100	359	167		17.12			
			Yes	100	603*	124		25.57*			
R-C5B			No	100	411	94		18.11			
			Yes	100	521*	193*		26.75*			
R-C5D			No	100	270	131		14.44			
			Yes	100	377*	384*		20.94*			

Table 3. *Continued*

Crop	Medium	Bacterium	Inoculation	Fertilizer N	Yield	Roots	N fixation	N uptake	^{15}N uptake	N content (%)	Reference
			<----	----	----	various units[a]	----	----	----	---->	
Zea mays	Soil	*A. brasilense*	No		1.470			10.23c		0.712	Nur et al. (1980)
		Cd	Yes		1.759			14.08a		0.801	
		Sp 7	Yes		1.538			11.24bc		0.731	
		Cd-1	Yes		1.828			13.56a		0.742	
		Cd-3	Yes		1.650			11.80bc		0.715	
Setaria italica			No		1.577b	0.4500b[c]	0	1.40b		0.906	
		Cd	Yes		2.821a	1.0020a	709	3.20a		1.127	
		Sp 7	Yes		2.569a	0.6842b	681	2.90ab		1.119	
		Cd-1	Yes		2.272ab	0.7856ab	585	2.50ab		1.084	
		Cd-3	Yes		2.175ab	0.5590b	557	2.10ab		0.988	
			No		1.4b		0	11.00b		0.80	
		Cd	Yes		2.5a		140	33.70a		1.33	
		Sp 7	Yes		2.4a		152	31.00a		1.23	
		Cd-1	Yes		2.2ab		140	20.00a		1.17	
		Cd-3	Yes		1.9ab		135	20.50ab		1.10	
Zea mays	Solution	*A. brasilense*	No	0	.59	0.27	0				Cohen et al. (1980)
		Sp 7	Yes	0	.85*	0.36	200				
			No	0.04	1.19	0.53	0				
			Yes	0.04	1.73*	0.76	156				
			No	0.08	2.09	0.86	0				
			Yes	0.08	2.81*	0.97	10				
			No	0.16	4.22	1.70	0				
			Yes	0.16	4.89	1.96	0				
			No	0.06	1.97	0.86	0				

Table 3. *Continued*

Crop	Medium	Bacterium	Inocu-lation	Ferti-lizer N	Yield	Roots	N fix-ation	N uptake	^{15}N uptake	N con-tent (%)	Reference
			<--- various units[a]							--->	
		Sp 7	Yes	0.06	3.05*	1.40	101				
		Sp-80	Yes	0.06	2.22	1.01	80				
		Cd	Yes	0.06	2.84*	1.35	99				
		Cd-1	Yes	0.06	2.21	0.98	77				
Setaria italica	Solution	*A. brasilense*	No	0.06	1.62	0.45	0	10.50		0.9	
		Sp 7	Yes	0.06	2.82*	1.23	112	15.90*		1	
		Sp-80	Yes	0.06	2.22	0.83	98	13.90		1	
		Cd	Yes	0.06	2.14	0.85	88	12.90		1	
		Cd-1	Yes	0.06	1.93	0.75	80	11.80		1	
Zea mays			No	0.06	4.53	1.60	0	20.80		0.71	
		Sp 7	Yes	0.06	5.72*	2.14	351	26.77*		0.75	
		Cd-1	Yes	0.06	5.46*	2.02	410	25.24*		0.73	
Setaria italica			No	0.06	1.19	0.54	0	5.22		0.8	
		Sp 7	Yes	0.06	2.15*	0.90	426	11.41*		0.91	
		Cd-1	Yes	0.06	2.04*	0.83	378	11.21*		0.93	
	Sand		No	0	0.87	0.53	0	2.20		0.65	
		Sp 7	Yes	0	1.55*	0.84	902	5.50*		0.78	
	Sandy Loam		No	17	1.74	0.71	0	6.70		0.65	
			Yes	17	3.05*	1.51	1322	12.90*		0.84	
	Loess		No	340	4.83	1.53	0	34.20		1.06	
			Yes	340	6.05*	2.05	2050	47.00*		1.19	

Table 3. *Continued*

Crop	Medium	Bacterium	Inoculation	Fertilizer N	Yield	Roots	N fixation	N uptake	^{15}N uptake	N content (%)	Reference
			<---------- various units[a]							---------->	
Zea mays	Solution	*A. brasilense*	No	0.26+ sucrose	310			1.39		0.46	Rennie (1980)
		Sp 7	Yes	0.26+ sucrose	403*			1.75*		0.43	
			No	0.26+ malate	333			1.55		0.46	
			Yes	0.26+ malate	470*			2.02*		0.43	
			No	0.26+ succinate	367			1.58		0.42	
			Yes	0.26+ succinate	580*			2.38*		0.41	
			No	0.26	383			2.11		0.55	
			Yes	0.26	453*			2.49*		0.55	
Zea mays	Soil	*A. brasilense*	No	0	12.4	2.52		49.4		0.4	Albrecht et al. (1981)
		Sp 7	Yes	0	11.3	3.67		59.9		0.52	
			No	15	15.7	4.27		156.0		0.99	
			Yes	15	14.7	4.68		140.0		0.95	
Digitaria sp	Clay	*A. brasilense*	No	0	15.9	3.6	3.4	0.221		1.4	Schank et al. (1981)
		(mixed)	Yes	0	17.3	4.3*	3.7	0.232		1.36	
	Sand		No	0	8.4	2.6	5.8	0.186		2.26	
			Yes	0	10.6*	3.1	5.4	0.196		1.85	
Panicum milaceum	Soil	*A. brasilense*	No		0.59	0.48		11.0		1.95	Kapulnik et al. (1981a)
		(mixed)	Yes		0.95*	1.10*		17.9*		1.88	
Sorghum bicolor × Sorghum sudanense			No		1.30	0.60		27.4		1.83	
			Yes		1.70*	1.32*		40.5*		2.28	

Table 3. *Continued*

Crop	Medium	Bacterium	Inoculation	Fertilizer N	Yield	Roots	N fixation	N uptake	^{15}N uptake	N content (%)	Reference
			<--- various units[a]							--->	
Triticum			No		0.58	0.54					
aestivum			Yes		1.09*	1.31*					
Sorghum			No		1.71	1.21	0	28.0		1.64	
bicolor			Yes		2.72**	1.18	61	48.4*		1.78	
Setaria	Soil	*A. brasilense*	No	0	0.93			10.69		1.15	Kapulnik et al.
italica		Cd	Yes	0	1.76*			18.48*		1.05	(1981b)
			No	0	0.79						
			Yes	0	0.98*						
			No	0.01	0.70						
			Yes	0.01	0.85*						
			No	0.04	0.83						
			Yes	0.04	1.25*						
			No	0.1	1.01						
			Yes	0.1	1.27*						
			No	0.2	1.08						
			Yes	0.2	1.29*						
Zea mays	Soil	*A. brasilense*	No	0.01	2.680c		8.8	14.16b		0.530	O'Hara et al.
Sweet September		Sp 7	Yes	0.01	3.955a		25.1	16.89abc		0.427	(1981)
		JM 6A2	Yes	0.01	3.267b		11.9	13.34c		0.408	
		Sp Br 14	Yes	0.01	3.560ab		53.3	16.05ab		0.445	
		Sp 107st	Yes	0.01	3.480b		3.9	16.14ab		0.462	
		Sp F104	Yes	0.01	3.650ab		0	17.02ab		0.463	
		Sp F105	Yes	0.01	3.822ab		3.5	17.28a		0.452	
		S-631	Yes	0.01	3.910a		46	15.47b		0.395	

Table 3. *Continued*

Crop	Medium	Bacterium	Inocu-lation	Ferti-lizer N	Yield	Roots	N fix-ation	N uptake	^{15}N uptake	N con-tent (%)	Reference
			<----	----	----	----	various units[a]	----	----	---->	
Zea mays			No	0.01	2.619d		2	12.87c		0.491	
John Innes		Sp 7	Yes	0.01	4.052a		25.1	22.30a		0.551	
		JM 6A2	Yes	0.01	2.992c		16.4	17.30b		0.591	
		Sp Br14	Yes	0.01	3.018c		0	15.75b		0.521	
		Sp 107st	Yes	0.01	3.558b		2.5	17.26b		0.485	
		Sp F104	Yes	0.01	3.095c		10.5	15.13b		0.488	
		Sp F105	Yes	0.01	3.692b		0	16.31b		0.441	
		S-631	Yes	0.01	3.304bc		21	17.47b		0.533	
Triticum durum	Soil	*A. brasilense*	No	Low	1.33					1.53	Avivi and Feldman (1982)
Responsive		(mixed)	Yes	Low	2.05**					1.49	
Nonresponsive			No	Low	1.82					1.50	
			Yes	Low	1.97					1.53	
Triticum aestivum			No	Low	1.14						
			Yes	Low	1.5*						
Triticum aestivum	Soil	*A. brasilense*	No	0	1.10	0.73	25.2			0.43	Hegazi (1982)
		+regulators	No	0	0.48	0.65	12.9			0.57	
		EgW2	Yes	0	1.10	0.83	177.0			0.41	
		WgG1	Yes	0	1.20	0.95	307.6			0.41	
		EgW2	Yes	500	1.89	0.65	39.2			1.19	
		WgG1	Yes	500	1.87	0.69	44.2			1.11	
			No	1000	1.73	0.59	16.5			1.77	

Table 3. *Continued*

Crop	Medium	Bacterium	Inoculation	Fertilizer N	Yield	Roots	N fixation	N uptake	^{15}N uptake	N content (%)	Reference
			<------------ various units[a] ------------>								
Pennisetum americanum	Soil	*A. brasilense*	No	0	4.20	1.62	4.57				Venkateswarlu and Rao (1983)
		S-7	Yes	0	4.93*	2.57*	5.18				
		S-14	Yes	0	5.95*	3.57*	4.17				
		S-51	Yes	0	4.08	1.67	4.62				
		S-54	Yes	0	5.36*	3.35*	5.57				
		S-59	Yes	0	4.60	2.35*	5.97				
	Solution		No	0.5m*M*	577	131	0.97	8.19		1.42	
		S-7	Yes	0.5m*M*	597	135	4.50	8.36		1.40	
		S-14	Yes	0.5m*M*	723*	187*	5.88	10.12*		1.40	
		S-51	Yes	0.5m*M*	633	136	5.95	8.99		1.42	
		S-54	Yes	0.5m*M*	700*	157	5.64	9.73*		1.39	
		S-59	Yes	0.5m*M*	610	137	3.65	8.72		1.43	
Zea mays	Soil	*A. lipoferum* (mixed)	No		11.4		26.3			0.71	Hegazi 5t al. (1983)
			Yes		22.9*		114.2			1.12	
			No	Straw	20.1		102.7			1.06	
			Yes	Straw	39.1*		144.7			1.39	
Zea mays	Soln	*A. brasilense*	No	N limiting	75.7						Lin et al. (1983)
		Cd	Yes	N limiting	95.9*						
		Sp 7	Yes	N limiting	101.7*						

Table 3. *Continued*

Crop	Medium	Bacterium	Inoculation	Fertilizer N	Yield	Roots	N fixation	N uptake	^{15}N uptake	N content (%)	Reference
			<------	------	------	------	various units[a]	------	------	------>	
Sorghum			No	N limiting	6.28						
bicolor		Cd	Yes	N limiting	7.28						
		Sp 7	Yes	N limiting	7.57*						
Triticum aestivum	Soil	*A. brasilense* Sp 107	No		712			3.76			Lethbridge and Davidson (1983)
			10ex9		640			3.76			
			10ex10		804*			4.95			
			10ex11		1024*			14.53			
			10ex11		1252*			17.24			
Rescue	Soil Arbuthnott 0.25% N		Dead	14	2.39			12.70	5.20		
			Live	14	2.43			13.30	5.46*		
C-R5D			Dead	14	2.32			13.20	5.34		
			Live	14	2.5*			13.40	5.64*		
R-C5D			Dead	14	2.35			13.70	5.61		
			Live	14	2.5*			13.40	5.47*		
Cadet			Dead	7	1.82			12.50	2.29		
			Live	7	1.89			13.10	2.38		
			Dead	14	2.19			13.10	5.18		
			Live	14	2.06*			13.00	5.26		
			Dead	28	2.31			12.70	10.50		
			Live	28	2.25			13.10	11.10*		
			Dead	56	2.67			11.80	22.80		
			Live	56	2.61			12.10	23.20*		
			Dead	112	3.64			10.70	47.90		
			Live	112	3.37*			11.50	47.30		

Table 3. *Continued*

Crop	Medium	Bacterium	Inoculation	Fertilizer N	Yield	Roots	N fixation	N uptake	^{15}N uptake	N content (%)	Reference
			<----------	----------	----------	various units[a]	----------	----------	----------	---------->	
	Soil Pittenheath		Dead	14	1.15			5.04	6.41		
	0.11% N		Live	14	1.23			5.12	6.49		
	Soil Countesswells		Dead	14	1.59			18.00	4.83		
	0.36% N		Live	14	1.51			18.00	4.32*		
Setaria italica	Soil	*A. brasilense*	No		2.87	0.53b		16.1b			Yahalom et al. (1984)
	Quartz sand		Yes		3.98*	0.87a		23.1a			
	Vermiculite		No		4.56	0.97b		27.3a			
			Yes		4.95	1.19a		28.8a			
	Rehovot		No		3.53			16.3b			
			Yes		4.24			20.4a			
Oryza sativa	Soil	*A. lipoferum*	No		2.64		300				Watanabe and Lin (1984)
			Yes		5.70*		500				
Triticum aestivum	Soil	*A. lipoferum*	No		62.0					1.5	Mertens and Hess (1984)
			Yes		71.9*					1.7	
			No		48.0					1.1	
			Yes		63.4					1.3	
Triticum aestivum	Soln	*A. brasilense*	No	0		3.65					Kapulnik et al. (1985c)
		Cd+Sp 7+Cd–1	Yes	0		4.43*					
	Vermiculite		No	0		11.55					
			Yes	0		14.30*					
	Soln		No	Hoag	3.61						
			Yes	Hoag	4.11						

Table 3. *Continued*

Crop	Medium	Bacterium	Inoculation	Fertilizer N	Yield	Roots	N fixation	N uptake	^{15}N uptake	N content (%)	Reference
			<-------------------------			various units[a]				------------------------->	
Lolium	Soil	*A. lipoferum*	No	40	17.4						Jagnow
perenne			Yes	40	17.8						(1985)
Hordeum			No	40	36						
vulgare			Yes	40	43						
Zea mays			No	40	648						
			Yes	40	632						
Lolium			No	40	2152						
perenne			Yes	40	2254						
Arrhenatherum			No	40	2016						
elatius			Yes	40	2306*						
Festuca			No	40	1384						
rubra			Yes	40	1593*						
Dactylis			No	40	3094						
glomerata			Yes	40	3246						
Vicia	Soil	*A. brasilense*	No	0	0.252	1.182				1.81	Sarig et al.
sativa		Cd/sp 7/ Cd–1	Yes	0	0.310*	0.209*				2.06*	(1986)
Pisum			No	0	0.192	0.069				1.3	
sativum			Yes	0	0.203	0.066				1.5	
Oryza sativa	Soil	*A. lipoferum*	No	180	35.7	5.5		639			Nayak et al.
0S4		34-H	Yes	180	39.5*	8.0*		765*			(1986)
Hua-chou-			No	180	40.4	2.6		718			
chi-mo			Yes	180	47.8*	4.7*		770			
Leptochloa	Soln	*A. brasilense*	No		80			0.88		1.09	Malik et al.
fusca			Yes		118*			1.43*		1.20	(1987)

Table 3. *Continued*

Crop	Medium	Bacterium	Inoculation	Fertilizer N	Yield	Roots	N fixation	N uptake	^{15}N uptake	N content (%)	Reference
			<------ various units[a]							------>	
Gossypium barbadense	Soil	*A. brasilense*	No		320.6		76.4	4.10			Fayez and Daw (1987)
		S-631	Yes		387.0*		183.4	5.01*			
		Sp Br14	Yes		382.8*		181.3	4.83*			
		E-15	Yes		388.9*		121.4	4.69*			
		M-4	Yes		340.5		71.2	4.46			
		SE	Yes		347.5		113.7	4.29			
		SC-22	Yes		354.3*		81.7	4.89*			
Zea mays Sweet September	Soil	*A. brasilense*	No	0	448.2a		8.8				O'Hara et al. (1987)
		Sp 7	Yes	0	418.0a		1.4				
		Sp F105	Yes	0	439.4a		1.4				
		Sp 107st	Yes	0	374.7a		8.4				
		Autoclaved culture		0	411.4a		0				
		Sterile medium		0	401.7a						
		Cell-free culture filtrate		0	465.6a						
		Washed bacterium suspension		0	468.2a						
			No	11.2	3281.6a		2				
		Sp 7	Yes	11.2	4280.0b		18.4				
		Sp F105	Yes	11.2	3816.0ab		3.5				
		Sp 107st	Yes	11.2	3745.2ab		4.2				
		Autoclaved culture		11.2	4078.8ab		0				
		Sterile medium		11.2	3846.3ab						
		Cell-free culture filtrate		11.2	3725.5ab						
		Washed bacterium suspension		11.2	4376.2b						

Table 3. *Continued*

Crop	Medium	Bacterium	Inoculation	Fertilizer N	Yield	Roots	N fixation	N uptake	^{15}N uptake	N content (%)	Reference
			<------ various units[a]	------	------	------	------	------	------	------>	
			No	22.4	9561.4a		0.5				
		Sp 7	Yes	22.4	9251.6a		0.5				
		Sp F105	Yes	22.4	9784.2a		1				
		Sp 107st	Yes	22.4	9814.7a		2				
		Autoclaved culture		22.4	9651.3a		0				
		Sterile medium		22.4	9521.6a						
		Cell-free culture filtrate		22.4	9410.5a						
		Washed bacterium suspension		22.4	9120.4a						
Pennisetum americanum (Gahi 3)			No	0	46.2a						
		Sp 7	Yes	0	51.0a						
		Sp F105	Yes	0	46.5a						
		Sp 107st	Yes	0	37.9a						
		Autoclaved culture		0	58.7a						
		Sterile medium		0	46.3a						
		Cell-free culture filtrate		0	42.8a						
		Washed bacterium suspension		0	49.5a						
			No	11.2	184.8ab						
		Sp 7	Yes	11.2	186.0ab						
		Sp F105	Yes	11.2	151.6a						
		Sp 107st	Yes	11.2	204.0b						
		Autoclaved culture		11.2	183.6ab						
		Sterile medium		11.2	185.2ab						
		Cell-free culture filtrate		11.2	187.4ab						

Table 3. *Continued*

Crop	Medium	Bacterium	Inoculation	Fertilizer N	Yield	Roots	N fixation	N uptake	^{15}N uptake	N content (%)	Reference
		<----------	various units[a]							---------->	
		Washed bacterium suspension		11.2	175.1a						
			No	22.4	284.3a						
		Sp 7	Yes	22.4	286.4a						
		Sp F105	Yes	22.4	305.7a						
		Sp 107st	Yes	22.4	298.4a						
		Autoclaved culture		22.4	304.2a						
		Sterile medium		22.4	200.7a						
		Cell-free culture filtrate		22.4	288.5a						
		Washed bacterium suspension		22.4	279.4a						
Zea mays Sweet September			No	8.4	1.9		8.6				
		Sp 7	Yes	8.4	2.2*		12.3				
		Sp 107st	Yes	8.4	2.0		15.5				
		Sp F105	Yes	8.4	2.3*		0.0				
			No	11.2	2.7		2.0				
		Sp 7	Yes	11.2	3.5*		18.4				
		Sp 107st	Yes	11.2	4.5*		4.2				
		Sp F105	Yes	11.2	3.6*		3.5				
Triticum aestivum (Expt. 1)	Solution	*A. brasilense*	No	1	0.26cde	0.18fg	0.00j				Ferreira et al. (1987)
		Sp 7	Yes	1	0.18h	0.14h	3.24b				
		Sp 107st	Yes	1	0.20gh	0.13h	3.61a				
		Sp F105	Yes	1	0.25e	0.22de	2.55c				
		M-3 nr^-	Yes	1	0.15i	0.10i	1.18e				

Table 3. *Continued*

Crop	Medium	Bacterium	Inoculation	Fertilizer N	Yield	Roots	N fixation	N uptake	^{15}N uptake	N content (%)	Reference
			<--- various units[a]							--->	
		M-15 nr⁻	Yes	1	0.20h	0.19efg	1.05efg				
		M-16 nr⁻	Yes	1	0.28bcde	0.23cd	0.45i				
		M-18 nr⁻	Yes	1	0.23f	0.18g	0.79fgh				
		M-22 nr⁻	Yes	1	0.22fg	0.15h	1.08ef				
			No	10	0.29b	0.23d	0.00j				
		Sp 7	Yes	10	0.29bc	0.22de	0.00j				
		Sp 107st	Yes	10	0.28bcd	0.22de	0.00j				
		Sp F105	Yes	10	0.37a	0.31a	0.00j				
		M-3 nr⁻	Yes	10	0.28bc	0.23cd	0.00j				
		M-15 nr⁻	Yes	10	0.29bc	0.26bc	1.53d				
		M-16 nr⁻	Yes	10	0.29b	0.28b	0.97efg				
		M-18 nr⁻	Yes	10	0.26de	0.21def	0.57hi				
		M-22 nr⁻	Yes	10	0.27bcde	0.19efg	0.74gh				
Triticum			No	1	0.23g	0.15g	0.00h				
aestivum		Sp 7	Yes	1	0.20h	0.15g	2.56c				
(Expt. 2)		Sp 107st	Yes	1	0.24fg	0.15g	4.44a				
		Sp F105	Yes	1	0.26de	0.21e	4.03b				
		M-15 nr⁻	Yes	1	0.23g	0.14g	1.56d				
		M-16 nr⁻	Yes	1	0.23g	0.20ef	1.18efg				
		M-18 nr⁻	Yes	1	0.23fg	0.18f	1.05fg				
			No	10	0.28bc	0.23cd	0.00h				
		Sp 7	Yes	10	0.25ef	0.21de	0.00h				
		Sp 107st	Yes	10	0.30b	0.24c	0.00h				
		Sp F105	Yes	10	0.38a	0.32a	0.00h				

Table 3. *Continued*

Crop	Medium	Bacterium	Inoculation	Fertilizer N	Yield	Roots	N fixation	N uptake	^{15}N uptake	N content (%)	Reference
			<----------	various units[a]						---------->	
		M-15 nr⁻	Yes	10	0.29b	0.27b	1.42de				
		M-16 nr⁻	Yes	10	0.27cd	0.29b	1.29def				
		M-18 nr⁻	Yes	10	0.29b	0.28b	0.97g				
						Number of Ears					
Triticum aestivum	Vermiculite		No	3	5.64d	0.6c		51.59c		1.23	
		Sp 245	Yes	3	6.06d	1c		57.25c		1.27	
		M-15 nr⁻	Yes	3	5.97d	0.8c		57.02c		1.28	
		M-16 nr⁻	Yes	3	6.08d	0.6c		55.38c		1.36	
		M-18 nr⁻	Yes	3	6.06d	0.8c		55.54c		1.32	
			No	10	8.04c	3.2cb		149.11b		2.79	
		Sp 245	Yes	10	8.86b	8.2a		160.73a		2.62	
		M-15 nr⁻	Yes	10	8.45bc	4.2b		164.05a		2.79	
		M-16 nr⁻	Yes	10	8.15c	4.2b		149.97b		2.80	
		M-18 nr⁻	Yes	10	9.44a	1.8c		158.58ab		2.81	
Triticum aestivum	Soil	*A. brasilense*	No	80	1161	1480		76.0			Warembourg et al. (1987)
		Sp Br14	Yes	80	1939*	3114*		91.3			
Triticum aestivum	Soil	*A. lipoferum*	No	0	4.7			108			Rai and Gaur (1988)
		Iso 1	Yes	0	6.7*			186*			
		Iso 2	Yes	0	6.4			159*			
		Iso 1+2	Yes	0	7.4*			210*			
			No	40	15.7			396			
		Iso 1	Yes	40	17.9*			468*			
		Iso 2	Yes	40	17.6*			423*			

Table 3. *Continued*

Crop	Medium	Bacterium	Inoculation	Fertilizer N	Yield	Roots	N fixation	N uptake	^{15}N uptake	N content (%)	Reference
			<----	----	----	----	various units[a]	----	----	---->	
		Iso 1+2	Yes	40	18.1*			537*			
			No	80	20.0			629			
		Iso 1	Yes	80	22.1*			686*			
		Iso 2	Yes	80	21.7			665*			
		Iso 1+2	Yes	80	23.1*			708*			
			No	120	25.7			789			
		Iso 1	Yes	120	25.8			781			
		Iso 2	Yes	120	25.7			754*			
		Iso 1+2	Yes	120	26.6			750*			
Oryza sativa	Solution	*A. lipoferum* 34H	No		0.47	0.33					Murty and Ladha (1988)
			Yes		0.51*	0.34					
Poa pratensis	Solution	*A. lipoferum* Fr	Dead		1.042			13.43		1.24	Haahtela et al. (1988)
			Live		0.945			10.55*		1.11	
Triticum aestivum			Dead		3.469			28.39		0.08	
			Live		2.986			23.86		0.08	
Lycopersicon esulentum	Soil	*A. brasilense* Cd	No		40	19					Hadas and Okon (1988)
			Yes		76*	28*					
Triticum aestivum	Soil	*A. lipoferum* 10	No	40	5.66			26.6			Darmwal and Gaur (1988)
			Yes	40	6.35*			30.8*			
			No	80	7.53			36.8			
			Yes	80	8.05*			39.9			
		No straw	No	40	6.85			136.0			
		No straw	Yes	40	7.80*			154.7*			

Table 3. *Continued*

Crop	Medium	Bacterium	Inoculation	Fertilizer N	Yield	Roots	N fixation	N uptake	^{15}N uptake	N content (%)	Reference
			<------	------	------	various units[a]	------	------	------	------>	
		No straw	No	80	9.65			193.8			
		No straw	Yes	80	10.53*			211.7*			
		Straw	No	40	6.09			108.7			
		Straw	Yes	40	6.58*			116.8			
		Straw	No	80	8.09			162.0			
		Straw	Yes	80	8.37*			159.8			

[a]Units vary among studies and are not reported because data are presented only to show relative increase due to inoculation.

[b]* = Significant at the 0.05 probability level.

[c]Values accompanied by the same letter in each column within a crop within a reference do not differ significantly at $P = 0.05$ level.

aestivum; Darmwal and Gaur (1988) with *Triticum aestivum;* Hegazi (1982) with *Sorghum bicolor, Zea mays,* and *Triticum aestivum;* Hegazi et al. (1983) with *Zea mays;* Ishac et al. (1986) with *Triticum aestivum;* Kapulnik et al. (1981c, 1985c, 1987) with *Zea mays, Setaria italica, Panicum miliaceum, Triticum aestivum,* and *Sorghum bicolor;* Mertens and Hess (1984) with *Triticum aestivum;* Raverkar and Konde (1988) with *Arachis hypogaea;* Reynders and Vlassak (1982a) with *Triticum aestivum;* Rai and Gaur (1982) with *Triticum aestivum;* Sarig et al. (1984, 1986, 1988) with *Sorghum bicolor, Cicer arietinum, Pisum sativum,* and *Vicia sativa;* Smith et al. (1984b) with *Sorghum bicolor* and *Pennisetum americanum;* Tilak and Subba Rao (1987) with *Pennisetum americanum;* Warembourg et al. (1987) with *Triticum aestivum;* and Yahalom et al. (1984) with *Setaria italica]*. With only one exception [Smith et al. (1984b) with *Sorghum bicolor]*, yield and N uptake or ARA increased concomitantly, indicating that at least part of the yield response is likely to be due to associative N fixation, but no major response to actual N_2 fixation has been demonstrated. Actual measurements of apparent N_2 fixation (ARA) in the field were not correlated with the N yield obtained (van Berkum and Bohlool, 1980; Kapulnik and Okon, 1983). However, the observed changes in root morphology and plant growth characteristics induced by inoculation are probably related to factors other than N_2 fixation. In addition, yield responses are often observed at moderate inorganic N levels (Table 4), suggesting that some of the N_2 fixed by *Azospirillum* becomes available to the plant. The inconsistency in yield responses to inoculation mentioned above is probably the result of ecological and environmental factors. The success of inoculation obviously depends on many factors, including the appropriate choice of carrier and inoculum, the ability of the bacterium to establish itself and to compete with the native microflora, favorable soil chemical and physical conditions such as pH, aeration, available nutrients including N, climatic conditions, and agricultural practices. In many of the early experiments, where responses were obtained to inoculation with the first strains of *Azospirillum* isolated, it is unlikely that the responses were due to N_2 fixation because the inoculated strain would have had to outcompete any native population for sites and have superior N_2-fixing ability as well. In cases where more than one strain of *Azospirillum* was used (Baldani et al., 1983; Boddey et al., 1986; Kapulnik et al., 1981c; Reynders and Vlassak, 1982a,b; Yahalom et al., 1984), the results strongly support the view that homologous isolates tend to be the most efficient at promoting yield and N uptake.

IV. Possible Reasons for Plant Response to *Azospirillum*

While there is little question that *Azospirillum* fixes nitrogen in the rhizosphere of plants, considerable controversy exists concerning the cause of the growth stimulation following inoculation. *Azospirillum brasilense* produces phytohormones which can stimulate root growth and induce changes in root morphology, which, in turn, could affect the assimilation of nutrients. These issues are now

Table 4. Yield responses of various crops to inoculation with *Azospirillum* spp. together with corresponding N content and uptake and acetylene reduction activity

Crop	Bacterium	Inoculation	Fertilizer[a] N (kg/ha)[a]	Yield Significance (kg/ha)[b]	N content (%)	N_2-ase	N uptake	R[c] or C	Reference and comments
Pennisetum americanum	*A. lipoferum*	No	0	5200					Smith et al. (1976)
		Yes	0	4800					
		No	20	6000					
		Yes	20	5900					
		No	40	6000					
		Yes	40	7100*					
		No	80	7500					
		Yes	80	9000*					
Panicum maximum, first harvest	*A. lipoferum*	No	0	4200					
		Yes	0	4000					
		No	20	5900					
		Yes	20	6500					
		No	40	6500					
		Yes	40	7300*					
		No	80	9000					
		Yes	80	9000					
Panicum maximum, first + second harvests	*A. lipoferum*	No	0	10100					
		Yes	0	9900					
		No	30	13000					
		Yes	30	15100*					
		No	60	15500					
		Yes	60	17000*					
		No	120	19000					
		Yes	120	20000					

Table 4. *Continued*

Crop	Bacterium	Inoculation	Fertilizer[a] N (kg/ha)[a]	Yield Significance (kg/ha)[b]	N content (%)	N_2-ase	N uptake	R[c] or C	Reference and comments
Pennisetum americanum	*S. lipoferum* Sp 13t	No	0	5200					Smith et al. (1977)
		Yes	0	4800					
		No	20	5800					
		Yes	20	6000					
		No	40	6000					
		Yes	40	7200*					
		No	80	7900					
		Yes	80	9100*					
Panicum maximum		No	0	4300					
		Yes	0	4000					
		No	20	5800					
		Yes	20	6400*					
		No	40	6500					
		Yes	40	7400*					
		No	80	9100					
		Yes	80	9200					
Pennisetum americanum	*Azospirillum* sp.	No	44.8	523					Taylor (1979)
		Yes	44.8	471					
		No	89.6	553					
		Yes	89.6	713*					
		No	134.4	598					
		Yes	134.4	601					

Table 4. *Continued*

Crop	Bacterium	Inoculation	Fertilizer[a] N (kg/ha)[a]	Yield Significance (kg/ha)[b]	N content (%)	N_2-ase	N uptake	R[c] or C	Reference and comments
Panicum maximum	*Azospirillum* sp.	No	44.8	1020					
		Yes	44.8	1002					
		No	89.6	1359					
		Yes	89.6	1621					
		No	134.4	1839					
		Yes	134.4	1919					
Pennisetum americanum	*A. brasilense*								Bouton et al. (1979)
Gahi 3	Sp 13t	Auto	60	10073					
		Live	60	11513*					
Tift 186		Auto	60	7482					
		Live	60	7188					
Tift 23DA		Auto	60	6258					
		Live	60	6169					
Bil 3D		Auto	60	5655					
		Live	60	6483*					
Tift 23DA		Auto	60	10900	1.74		191	R	
× Tift 18DB		Live	60	11234	1.73		194	R	
Tift 23DA		Auto	60	19588	1.43		280	R	
× Tiflate		Live	60	17858	1.58		282	R	
Tift 23DA		Auto	60	11523	2.01		231	R	
× Tift 383		Live	60	10712	2.02		215	R	
Tift 23DA		Auto	60	13116	1.50		198	R	
× Tift 186		Live	60	17269*	1.57		272	R	

Table 4. *Continued*

Crop	Bacterium	Inocu-lation	Fertilizer[a] N (kg/ha)[a]	Yield Signifi-cance (kg/ha)[b]	N content (%)	N_2-ase	N uptake	R[c] or C	Reference and comments
Tift 23DA		Auto	60	11791	1.65		195	R	
× PMP 004		Live	60	11241	1.62		180	R	
Tift 23A		Auto	60	8978	1.68		151	R	
× Bil 3B		Live	60	9229	1.60		151	R	
Tift 23DA		Auto	60	7647	1.89		145	R	
		Live	60	7529	2.10		158	R	
Tift 23DB		Auto	60	7018	1.94		136	R	
		Live	60	6168	2.01		124	R	
Tift 23A		Auto	60	8269	1.71		141	R	
		Live	60	7927	1.78		141	R	
Tift 23D		Auto	60	8961	1.59		141	R	
		Live	60	8545	1.65		140	R	
Tift 239A		Auto	60	6726	1.91		128	R	
		Live	60	6777	1.90		129	R	
Tift 239B		Auto	60	6519	1.75		115	R	
		Live	60	7354	1.83		134	R	
Tift 186		Auto	60	11363	1.97		224	R	
		Live	60	10368	1.91		198	R	
Tift 383		Auto	60	7400	2.05		153	R	
		Live	60	8030	1.96		155	R	
Tiftlate		Auto	60	14861	1.48		220	R	
		Live	60	16012	1.55		248	R	
Tift 23SB		Auto	60	10113	1.65		168	R	
		Live	60	9527	1.63		155	R	

Table 4. *Continued*

Crop	Bacterium	Inoculation	Fertilizer[a] N (kg/ha)[a]	Yield Significance (kg/ha)[b]	N content (%)	N_2-ase	N uptake	R[c] or C	Reference and comments
Bil 3B		Auto	60	6831	1.53		105	R	
		Live	60	8023	1.47		118	R	
Tift 13		Auto	60	8738	1.54		134	R	
		Live	60	7859	1.52		117	R	
Tift 131		Auto	60	8749	1.80		155	R	
		Live	60	9271	1.59		148	R	
Tift 123		Auto	60	9856	1.91		187	R	
		Live	60	10050	1.89		187	R	
Tift 18DB		Auto	60	6457	2.07		133	R	
		Live	60	6001	2.24		134	R	
Sorghum	*A. brasilense*	No	0	1167	0.65		63.5	R	Pal and Malik
bicolor		Yes	0	1780	0.66		78.2	R	(1981)
CSH-5 in 1978		No	40 B	1557	0.73		68.0	R	
		Yes	40 B	2810*	0.74		86.8	R	
		No	40 T	2796	0.73		86.2	R	
		Yes	40 T	2727	0.74		87.6	R	
		No	FYM	2048	0.67		66.5	R	
		Yes	FYM	2435	0.70		83.0	R	
CSH-5 in 1979		No	0	2427	0.86		95.2	R	
		Yes	0	2818	1.00		101.6	R	
		No	40 B	3250	0.99		102.5	R	
		Yes	40 B	3085	1.01		107.8	R	
		No	40 T	3414	0.98		103.5	R	
		Yes	40 T	2982	1.02		108.3	R	
		No	FYM	2756	0.86		98.9	R	
		Yes	FYM	2622	1.00		104.6	R	

Table 4. *Continued*

Crop	Bacterium	Inoculation	Fertilizer[a] N (kg/ha)[a]	Yield Significance (kg/ha)[b]	N content (%)	N_2-ase	N uptake	R[c] or C	Reference and comments
Zea mays	*A. brasilense*								Albrecht et al. (1981)
Inbred lines	Sp 7	No	0	11333	1.16		130	R	
		Yes	0	11778	1.21		142	R	
Sweet		No	0	18500	1.06		196	R	
		Yes	0	20389	1.11		226	R	
Inbred lines		No	20	5556	1.61		89	R	
		Yes	20	5778	1.11		64	R	
Hybrid		No	20	7667	1.32		98	R	
		Yes	20	7222	1.22		85	R	
Inbred lines	Mixed	No	0	3500	1.54		57	R	
		Yes	0	3000	1.57		46	R	
Hybrid		No	0	6778	1.44		97	R	
		Yes	0	5000	1.42		77	R	
Inbred lines		No	0	6278	0.78		49	R	
		Yes	0	7722*	0.82		63	R	
Hybrid		No	0	8000	0.64		51	R	
		Yes	0	9944*	0.78		78	R	
				ears					
Zea mays	*A. brasilense*	No	0	18620		0			Kapulnik et al. (1981c)
Sweet	Cd	Yes	0	20170		85			
		No	60	21550		0			
		Yes	60	23330		170			
		No	120	24620		0			
		Yes	120	28000*		105			
		No	240	22930		0			
		Yes	240	25730		15			

Table 4. *Continued*

Crop	Bacterium	Inoculation	Fertilizer[a] N (kg/ha)[a]	Yield Significance (kg/ha)[b]	N content (%)	N_2-ase	N uptake	R[c] or C	Reference and comments
				tops					
Zea mays	*A. brasilense*	No		11930			103.8	R	
		Yes		14130*			180.9	R	
Sorghum bicolor	*A. brasilense*	No		2880	1.52		43.78	C	
	Cd	Yes		3890*	1.67		64.96	C	
	Cd-1	Yes		3470	1.57		54.48	C	
Setaria italica	*A. brasilense*	No		1442	0.65		11.48	R	
		Yes		2092*	1.12		26.20	R	
Zea mays	*A. brasilense*	No		7420					
	Cd	Yes		8200*					
	Cd-1	Yes		7910					
Panicum miliaceum	*A. brasilense*	No		2798					
		Yes		3168*					
Triticum aestivum	*A. brasilense*	No		6411			144	R	Reynders and Vlassak (1982a)
	S-631	Yes		6997*			177	R	
	Sp Br14	Yes		7361*			183	R	
	A. brasilense	No	0	3328					
		Yes	0	3387					
		No	80	4200					
		Yes	80	4189					
Triticum aestivum	*A. Lipoferum* 20	No	0	1260	1.97		33.0	R	Rai and Gaur (1982)
		Yes	0	2070*	2.06		57.6	R	
		No	40	2370	2.10		69.9	R	
		Yes	40	3110*	2.10		87.4	R	
		No	80	2960	2.06		79.9	R	
		Yes	80	4150*	2.06		111.9	R	

Table 4. *Continued*

Crop	Bacterium	Inocu-lation	Fertilizer[a] N (kg/ha)[a]	Yield Signifi-cance (kg/ha)[b]	N content (%)	N_2-ase	N uptake	R[c] or C	Reference and comments
Sorghum bicolor	*A. lipo/brasil*	No	0	1194	1.71	94	20.42	C	Hegazi (1982)
Giza 3		Yes	0	1918*	1.94	338	37.21	C	
		No	200	1918	3.19	55	61.18	C	
		Yes	50	2072	2.72	486	56.36	C	
Giza 15		No	0	1570	1.70	81	26.69	C	
		Yes	0	2812*	1.74	560	48.93	C	
		No	200	3342	2.86	28	95.58	C	
		Yes	50	3747*	1.90	608	71.19	C	
Giza 114		No	0	1327	1.82	122	24.15	C	
		Yes	0	2780*	1.93	892	53.65	C	
		No	200	2710	3.01	49	81.57	C	
		Yes	50	3565*	2.19	391	78.07	C	
Shnab	*A. lipo/brasil*	No	0	474	1.80	1.2	20.31	C	
		Yes	0	2236*	1.91	14.5	101.67	C	
		No	Yes	2138	2.00	3.0	101.81	C	
Giza 156		No	0	798	1.70	4.7	32.28	C	
		Yes	0	3214*	1.80	10.6	137.75	C	
		No	Yes	2398	2.04	1.0	116.46	C	
Triticum aestivum									
Inbar	*A. brasilense*	No		4220					Kapulnik et al. (1983)
	Cd	Yes		4440					
	Cd-1	Yes		4680*					
Barkai		No	120	3389					
	Cd and Cd-1	Yes	120	3660*					

Table 4. *Continued*

Crop	Bacterium	Inoculation	Fertilizer[a] N (kg/ha)[a]	Yield Significance (kg/ha)[b]	N content (%)	N_2-ase	N uptake	R[c] or C	Reference and comments
Miriam		No	0	3590					
		Yes	0	3849*					
		No	40	3810					
		Yes	40	4010*					
		No	80	3920					
		Yes	80	3990					
Oryza sativa	*Azospirillum* sp.	No	0	2930					Rao et al. (1983)
		Yes	0	3080*					
		No	30	3520					
		Yes	30	3980*					
		No	45	3930					
		Yes	45	4540*					
		No	60	4450					
		Yes	60	4850*					
Oryza sativa	*Azospirillum* sp.	No	90	2550					Kannaiyan et al. (1983)
		Yes	90	3200(?)					
		No	120	2230					
		Yes	120	3200(?)					
Zea mays									
Galal	*A. lipoferum*	No	0	3218	1.42	288	86.00		Hegazi et al. (1983)
		Yes	0	8171*	1.75	839	213.00		
		No	200	8698	1.82	28	272.00		
Double cross		No	0	3347	1.13		71.60		
		Yes	0	3688	1.34		101.50		
		No	200	8237	1.75		212.25		

Table 4. *Continued*

Crop	Bacterium	Inoculation	Fertilizer[a] N (kg/ha)[a]	Yield Significance (kg/ha)[b]	N content (%)	N_2-ase	N uptake	R[c] or C	Reference and comments
American early		No	0	2648	1.28		73.80		
		Yes	0	4829*	1.68		170.75		
		No	200	8247	1.47		226.41		
Triticum aestivum	*A. brasilense*	No	30	0.85			14.0	R	Baldani et al.
	Sp Br14	Yes	30	0.88			14.5	R	(1983),
	Sp 107st	Yes	30	0.95			17.0	R	yield in g/plant,
	Sp 242st	Yes	30	0.90			15.0	R	N uptake in
	Sp 245st	Yes	30	1.05*			22.0	R	mg/plant
Sorghum bicolor	*A. brasilense*	Killed	30	5385	1.35	64.7	72.7	C	Smith et al.
Funk 552	Cd	Live	30	5951*	1.35	55.3	179.1	C	(1984b)
Pennisetum americanum		Killed	30	10848	1.26	36	136.7	C	
Gahi-3		Live	30	10733	1.18	40	126.6	C	
Pennisetum sp.	Cd+CdSR	Killed	0	8000			69.5	R	
hybrid	Cd	Live	0	8550			80.0	R	
	CdSR	Live	0	9000			75.0	R	
	Cd+CdSR	Killed	30	7250			59.0	R	
	Cd	Live	30	9020*			80.0	R	
	CdSR	Live	30	8500			78.0	R	
	Cd+CdSR	Killed	60	8930			72.0	R	
	Cd	Live	60	10500*			92.0	R	
	CdSR	Live	60	9200			79.0	R	
	Cd+CdSR	Killed	120	10900			90.0	R	
	Cd	Live	120	10700			95.0	R	
	CdSR	Live	120	10300			94.0	R	

Table 4. *Continued*

Crop	Bacterium	Inocu-lation	Fertilizer[a] N (kg/ha)[a]	Yield Signifi-cance (kg/ha)[b]	N content (%)	N_2-ase	N uptake	R[c] or C	Reference and comments
Triticum durum	*A. brasilense*	No	0	16	1.70				Zambre et al. (1984), yield reported in g/ha (sic)
		Yes	0	18	1.78				
		No	30	21	1.87				
		Yes	30	25	1.97				
		No	60	25	1.99				
		Yes	60	31	2.09				
		No	90	29	2.18				
		Yes	90	37	2.24				
		No	120	34	2.23				
		Yes	120	42	2.28				
Setaria italica	*A. brasilense*	No	70	1421	2.32		27.9	R	Yahalom et al. (1984)
	Cd	Yes	70	2244*	2.00		45.1	R	
	Sp 7	Yes	70	1954	2.05		39.1	R	
Sorghum bicolor	*A. brasilense*	No	57	9480	0.95		90.0	R	Sarig et al. (1984)
Hay		Yes	57	11270*	1.52		171.3	R	
Seed		No	57	3385					
		Yes	57	3989*					
Triticum aestivum	*A. lipoferum*	No	0	300	1.20		3.60	C	Mertens and Hess (1984)
		Yes	0	510*	1.20		6.12	C	
		No	60	980	1.20		11.76	C	
		Yes	60	1340*					

Table 4. *Continued*

Crop	Bacterium	Inocu-lation	Fertilizer[a] N (kg/ha)[a]	Yield Signifi-cance (kg/ha)[b]	N content (%)	N_2-ase	N uptake	R[c] or C	Reference and comments
Triticum aestivum	*A. brasilense*	No	0	234	0.44		7.60	R	Kapulnik et al.
Hay		Yes	0	211	0.36		10.29	R	(1985b)
		No	37	490	0.44		21.56	R	
		Yes	37	496	0.42		20.80	R	
		No	74	725	0.80		58.00	R	
		Yes	74	782	0.75		58.65	R	
		No	112	453	1.56		70.67	R	
		Yes	112	773*	1.05		81.10	R	
Pennisetum americanum		No	0	830					Wani et al.
IP 2787	*A. lipoferum*	Yes	0	920					(1985)
	A. brasilense Sp 7	Yes	0	760					
ICMS 7819		No	0	1310					
	A. lipoferum	Yes	0	1360					
	A. brasilense Sp 7	Yes	0	1240					
Mean		No	0	1860	1.23		37.0	R	
	A. lipoferum 1	Yes	0	2130	1.25		41.3	R	
	A. lipoferum ICM 1001	Yes	0	2250	1.27		46.8	R	
	A. brasilense Sp 7	Yes	0	1640	1.26		33.3	R	
Mean	*A. lipoferum*	No	20	2730	1.40				
	Strain 1	Yes	20	2870	1.40				
	Strain ICM 1001	Yes	20	2890	1.48				
		No	16	1460	0.87				
	Strain 1	Yes	16	1470	0.85				
	Strain ICM 1001	Yes	16	1540	0.81				

Table 4. *Continued*

Crop	Bacterium	Inoculation	Fertilizer[a] N (kg/ha)[a]	Yield Significance (kg/ha)[b]	N content (%)	N_2-ase	N uptake	R[c] or C	Reference and comments
Triticum aestivum									
West-breed	*A. brasilense*	No	67	1609					Hegazi and Saleh (1985)
		Yes	67	2603*					
Yoco-Rogo		No	67	861					
		Yes	67	902					
Oryza sativa									
Cesariot	*A. lipoferum* 4B	Killed	120	3760					Charyulu et al. (1985)
		Yes	120	4460***					
M-17		Killed	120	2280					
		Yes	120	3260***					
Cesariot		No	0	2260					
		Yes	0	4500***					
M-17		No	0	3390					
		Yes	0	4310**					
Cesariot		No	92	4910					
		Yes	92	3310*					
M-17		No	92	5080					
		Yes	92	3640*					
Triticum aestivum	*A. brasilense*	No	100	3951					Millet et al. (1985)
	Sp 7/CD	Yes	100	4870					
TAA64		No	100	6340					
Mivhor location	Sp 7/CD/new	Yes	100	6230					
	Sp 7/CD	Yes	100	6440					
	New 22b/22c/23	Yes	100	6470					
	New 4/5/6/8	Yes	100	6070					

Table 4. *Continued*

Crop	Bacterium	Inoculation	Fertilizer[a] N (kg/ha)[a]	Yield Significance (kg/ha)[b]	N content (%)	N_2-ase	N uptake	R[c] or C	Reference and comments
TAA66		No	100	6520					
Mivhor location	Sp 7/CD/new	Yes	100	6450					
	Sp 7/CD	Yes	100	6400					
	New 22b/22c/23	Yes	100	6480					
	New 4/5/6/8	Yes	100	6960*					
Triticum turgidum		No	100	7160					
TTR38	Sp 7/CD/new	Yes	100	7160					
Mivhor location	Sp 7/CD	Yes	100	7100					
	New 22b/22c/23	Yes	100	6600*					
	New 4/5/6/8	Yes	100	7320					
Triticum aestivum		No	100	3460					
TAA64	Sp 7/CD/new	Yes	100	3960					
Nir-Am Location	Sp 7/CD	Yes	100	3930					
	New 22b/22c/23	Yes	100	3700					
TAA66		No	100	4200					
Nir-Am Location	Sp 7/CD/new	Yes	100	3890					
	Sp 7/CD	Yes	100	4130					
	New 22b/22c/23	Yes	100	4100					
Triticum turgidum		No	100	5190					
TTR38	Sp 7/CD/new	Yes	100	5310					
	Sp 7/CD	Yes	100	5160					
	New 22b/22c/23	Yes	100	5360					

Table 4. *Continued*

Crop	Bacterium	Inoculation	Fertilizer[a] N (kg/ha)[a]	Yield Significance (kg/ha)[b]	N content (%)	N_2-ase	N uptake	R[c] or C	Reference and comments
Triticum aestivum	*A. lipo/brasil*	No	0	5325		4.29			Ishac et al. (1986)
		Yes	0	7325**		4.69			
		No	75	8000		4.19			
		Yes	75	7750		4.5			
		No	150	8500		5.48			
		Yes	150	8750		5.26			
Triticum aestivum	*A. brasilense*	Dead	30	232	2.52		5.9	R	Boddey et al. (1986)
	Sp 245	Yes	30	415*	2.98		12.4	R	
	Sp 245 nr⁻	Yes	30	256	2.69		6.9	R	
	Sp 246	Yes	30	351*	2.94		10.3	R	
	Sp 7	Yes	30	270	2.79		7.5	R	
	A. amazonense YTr	Yes	30	319*	2.95		9.4	R	
Cicer arietinum	*A. brasilense*	No	0	2860					Sarig et al. (1986), *Pisum sativum* yield in g/plant
Chick pea	Mixed	Yes	0	3540*					
Pisum sativum		No	0	16					
		Yes	0	17					
Vicia sativa		No	0	6760	2.00	0.38			
		Yes	0	8340*	2.33	0.81			
Hedysarum coronarium		No	0	6150		5.36			
		Yes	0	6780		9.31			

Table 4. *Continued*

Crop	Bacterium	Inoculation	Fertilizer[a] N (kg/ha)[a]	Yield Significance (kg/ha)[b]	N content (%)	N_2-ase	N uptake	R[c] or C	Reference and comments
		Soil—Kanpur pH 7.5							
Pennisetum americanum	*A. brasilense*	No	0	1275			60.50	R	Tilak and Subba Rao (1987)
		Yes	0	1550*			75.75	R	
		No	10	1625			86.15	R	
		Yes	10	1910*			106.50	R	
		No	20	1750			117.15	R	
		Yes	20	1875*			138.57	R	
		No	40	2012			162.75	R	
		Yes	40	2050			180.60	R	
		Soil-Hyderabad pH 6.8							
		No	0	1250					
		Yes	0	1275					
		No	10	1425					
		Yes	10	1450					
		No	20	1575					
		Yes	20	1695					
		No	40	1850					
		Yes	40	1925					
		Soil-Parbhani pH 7.2							
		No	0	1450					
		Yes	0	1765*					
		No	10	1950					
		Yes	10	2275*					
		No	20	2250					
		Yes	20	2315					
		No	40	2425					
		Yes	40	2610					

Table 4. *Continued*

Crop	Bacterium	Inoculation	Fertilizer[a] N (kg/ha)[a]	Yield Significance (kg/ha)[b]	N content (%)	N_2-ase	N uptake	R[c] or C	Reference and comments
		Soil—Dehli pH 7.8							
		No	0	975					
		Yes	0	1125					
		No	10	1175					
		Yes	10	1480*					
		No	20	1250					
		Yes	20	1575*					
		No	40	1850					
		Yes	40	2075					
Triticum aestivum	*A. brasilense*	No		3009	1.81		55	R	Warembourg et al.
	Sp Br 14	Yes		4640*	1.77		82	R	(1987)
Triticum aestivum	*A. brasilense*	No	15	2292	2.48		52.75	R	Baldani et al.
	Sp 245	Yes	15	2254	2.66		63.25	R	(1987)
	Sp 107st	Yes	15	2292	2.67		61.50	R	
	Sp 7 + Cd	Yes	15	2261	2.27		47.25	R	
		No	60	2284	2.50		62.00	R	
	Sp 245	Yes	60	2351	2.60		59.25	R	
	Sp 107st	Yes	60	2284	2.69		56.50	R	
	Sp 7 + Cd	Yes	60	2261	2.49		57.25	R	
		No	15	696	2.27		159.25	R	
		No	100	712	2.40		169.75	R	
	Sp 245	Yes	15	912	2.47		225.00	R	
	Sp 245 NR	Yes	15	846	2.27		192.75	R	
	Sp Trl	Yes	15	586	3.32		136.00	R	
	A. amazonense	YTrl	15	815	2.34		190.00	R	

Table 4. *Continued*

Crop	Bacterium	Inoculation	Fertilizer[a] N (kg/ha)[a]	Yield Significance (kg/ha)[b]	N content (%)	N_2-ase	N uptake	R[c] or C	Reference and comments
Triticum aestivum	*A. brasilense* mixed	No	0	2250					Kapulnik et al. (1987)
		Yes	0	2830*					
		No	60	4080					
		Yes	60	4340					
		No	120	4320					
		Yes	120	4900*					
		No	0	3990					
		Yes	0	4390*					
		No	40	4610					
		Yes	40	5100*					
		No	80	5210					
		Yes	80	5440					
Arachis hypogaea Robut 33-1	*A. lipoferum*	No	0	2210	2.35		82	R	Raverkar and Konde (1988)
		Yes	0	2730*	3.02		138	R	
	+ *Rhizobium* NC92	No	0	2790	3.78		194	R	
		Yes	0	2280*	2.63		109	R	
JL-24	*A. lipoferum*	No	0	2230	2.21		80	R	
		Yes	0	2780*	2.44		149	R	
	+ *Rhizobium* NC92	No	0	2840	3.58		190	R	
		Yes	0	2280*	2.26		102	R	

Table 4. *Continued*

Crop	Bacterium	Inoculation	Fertilizer[a] N (kg/ha)[a]	Yield Significance (kg/ha)[b]	N content (%)	N_2-ase	N uptake	R^c or C	Reference and comments
Zea mays	*Azospirillum*	No	90	13300					Okon et al.
Sweet		Yes	90	13800					(1988)
Sorghum bicolor	*A. brasilense*	No	0	4400					Sarig et al.
		Yes	0	5100					(1988)
		No	0	2500					
		Yes	0	2800					
		No	0	2700					
		Yes	0	3200					
Zea mays	*Azospirillum*		0	18.4*					Okon et al.
Forage			100	17.0*					(1988)
			50	13.7*					
Sweet (ears)			50	29.6*					
			50	26.5*					
			65	0.0					
			100	13.5					
			25	38.0*					
			50	4.9					
			50	6.8					
			Organic	10.5					
Field			Organic	16.5					
			0	15.2*					

Table 4. *Continued*

Crop	Bacterium	Inocu-lation	Fertilizer[a] N (kg/ha)[a]	Yield Signifi-cance (kg/ha)[b]	N content (%)	N_2-ase	N uptake	R[c] or C	Reference and comments
T. aestivum			0	19.5*					
Forage Grain			33	5.2*					
			100	7.9*					
			66	9.0					
			33	2.6					
			100	13.4*					
			75	6.0					
			0	3.1					
			0	12.0*					
			Organic	10.9					
Sorghum bicolor			0	18.8*					
Forage Grain			0	12.1*					
			0	17.6*					
			0	19.9*					
			0	15.0*					
Setaria italica			0	47.5*					
Forage			0	38.5*					
Panicum miliaceum			Organic	13.2*					
Seeds									

[a]B = Basal dressing. T = Top dressing at 30 days after planting. FYM = Farm yard manure.

[b]* = Significant at the 0.05 level. ** = Significant at the 0.01 level.

[a]% increase of yield from noninoculated controls.

[b]Level of N fertilization (% recommended).

[c]Reported or calculated.

discussed so that a clearer picture of the reasons for growth stimulation may be obtained.

A. Nutritional Effects

1. N_2 Fixation

For N_2 to be fixed by *Azospirillum*, adequate levels of substrate in the rhizosphere must be present. Much of this substrate would come from plant exudates. A recurrent question, which has not been answered satisfactorily to date, is whether these exudates are present in sufficient quantities to support significant levels of growth and N_2 fixation. Present techniques do not appear to be able to resolve this question (Hubbell and Gaskins, 1984). Many attempts have been made to measure and quantify the amount of N_2 fixed and transferred to the plant by *Azospirillum* based on N uptake and ARA assays. In some cases these efforts have been quite successful, based on ARA measurements on undisturbed *Setaria italica* plants inoculated with *A. brasilense* strain Cd (Cohen et al., 1980; Kapulnik et al., 1981b; Yahalom et al., 1984) and in sorghum and wheat based on total N yield from soils very low in N (Mertens and Hess, 1984; Sarig et al., 1984) as well as under conditions unfavorable to N_2 fixation, such as low soil temperatures or high levels of combined N with maize, sorghum, forage grasses, rice, and wheat (Avivi and Feldman, 1982; Kapulnik et al. 1981a,b, 1982, 1983; Mertens and Hess, 1984; Millet et al., 1985; Pal and Malik, 1981; Rai and Gaur, 1982, 1988; Reynders and Vlassak, 1982a, 1982b; Smith et al., 1984a, 1984b; Yahalom et al., 1984). However, there are many others where the technique used in the ARA assay did not allow extrapolation to actual N_2 fixed, or the results did not correlate with the N yield responses obtained (Bouton and Zuberer, 1979; Bouton et al., 1979; Hegazi et al., 1979, 1983, 1985; Kapulnik et al., 1982; O'Hara et al., 1981; Schank et al., 1981). Lethbridge and Davidson (1983) even suggested that wheat plants would only derive minor amounts of N from *Azospirillum* if exogenous carbohydrates were not present in the rhizosphere. Rao et al. (1987) have suggested that plant types with efficient photosynthate excretion through the roots should be selected to enhance the benefit from *Azospirillum* inoculation. On the other hand, Warembourg et al. (1987) showed that the respiratory activity in the rhizosphere of inoculated plants did not differ from that of the controls on a unit root weight basis, indicating that bacterial inoculum which greatly affected root growth did not require more energy in the form of exudates. Although N yield was greatly increased by inoculation, N contents in grain and tops were lower in inoculated than in control plants. Warembourg et al. (1987) resorted to unnecessarily long arguments to explain this phenomenon, but it would appear that inoculated plants simply were accumulating C relative to N at a faster rate than the controls, which would be required to account for the increased yields observed. This phenomenon is quite common in plant nutrition. Because the ARA measured on undisturbed plants has been of the order required to account for the increase in N yield of the plants in some cases, it has been tempting to attribute inoculation responses to biological N_2 fixation. However, in other studies (Lethbridge et al., 1982; O'Hara et al., 1981, 1987) the levels of N fixed by the bacteria were too low to contribute significantly to the N requirements of the plants. Okon et al. (1983), using ^{15}N-labeled N_2 and ARA assay, demonstrated that *A. brasilense* in the rhizosphere of *Setaria italica* fixed N_2, but only about 5% of the N_2 fixed was incorporated into the plant. The N_2 was transferred only very slowly, probably requiring bacterial decomposition and mineralization. These results are in agreement with the work on rice, which shows that only about 10 to 25% of the ^{15}N fixed in the root zone was transferred to the plant (Rao et al., 1987). Much of the N_2 fixed in a rice ecosystem could be due to heterotrophic organisms, which could also be true to a lesser extent in other nonsterile soils used in some of the above experiments. In addition, Kapulnik et al. (1985b) and Boddey et al. (1986), using ^{15}N-labeled fertilizer and ARA measurements, showed that the increases in grain yield and N content in Table 4 were not due to N_2 fixation but rather to increased nitrate assimilation arising from the increased nitrate reductase activity of the *Azospirillum* in the roots. These findings are supported by the work of Okon (1985) and Okon and Kapulnik (1986), who showed that the biological nitrogen fixation due to inoculation with *Azospirillum* was very small and insufficient to contribute substantial amounts of fixed N to the plant. Extrapolating experimental data, Okon et al. (1988a) suggest that less than 1 kg/ha per season is supplied in most cases by *Azospirillum*. On the other hand, Malik et al. (1987) showed that using *A. brasilense* strain DSM 1691 and ^{15}N, the percentage of N in kallar grass derived from atmospheric N was between 30 and 60%. Lima et al. (1987) demonstrated that up to 50% of the N content of crops such as sugarcane, *Panicum maximum*, and *Paspalum notatum* could be supplied by bacterial–plant-associated biological N fixation in which *Azospirillum* may function among other N_2-fixing species. Nayak et al. (1986) concluded on the basis of ^{15}N dilution and ARA techniques that growth promotion and increased N uptake in rice inoculated with *A. lipoferum* could not be accounted for by higher N_2 fixation. Ferreira et al. (1987) also demonstrated a role for bacterial nitrate reductase in the nitrate metabolism of the plant, which was probably responsible for the enhanced yields of plants inoculated with *A. brasilense* strain Sp 245 against its nitrate reductase negative mutants. These mutants showed no increase in shoot or root weights. They showed that a comparison of the behavior of strain Sp 245 and its NR^- mutants on root growth and ARA pointed to an increased flux of nitrate into the root, possibly due to changed root structure caused by hormones from the bacteria or increased ion uptake. An additional effect over and above that on roots appears as pronounced differences in nitrate reductase activity in shoots between strain Sp 245 and its nitrate reductase negative mutants. The bacterial nitrate reductase enhances nitrate reduction in the roots, and the parent strain Sp 245 seems to supply more reduced N to the shoots, resulting in increased yield. In a field experiment comparing strain Sp 245 and its nitrate reductase negative mutant, the latter was found to be far less effective than the parent strain in penetrating the interior of the root and promoting yield and N uptake in wheat

(Baldani et al., 1986; Boddey et al., 1986). Kapulnik et al. (1987) have demonstrated that ARA on inoculated wheat plants occurred only at the lowest N levels (whereas the *Azospirillum* effect was observed at all N levels) and then only after spike emergence, which leads to the conclusion that the main effect of *Azospirillum* inoculation occurs prior to maximum N_2 fixation. Any N_2 fixed at this late stage of growth would have little impact on yield, which supports the view previously expressed (Boddey and Dobereiner, 1982; Kapulnik et al., 1985b; Lethbridge and Davidson, 1983; Mertens and Hess, 1984) that *Azospirillum* inoculation does not contribute significantly to improved growth and yield of wheat plants through N_2 fixation. Bouton (1988) has shown that the ARA of pearl millet roots inoculated with *A. brasilense* strain JM 125 was only 2.5% of that for alfalfa grown on the same soil. In fact, there have been no reports in the literature in which *Azospirillum* inoculation responses can be definitely attributed to biological N_2 fixation alone.

The great complexity involved in the observed yield responses to *Azospirillum* inoculation is illustrated in the work of O'Hara et al. (1987), who set out to definitively isolate the causes using various plants, strains (nitrate respiring [nir^-], denitrifying [nir^+] and non-N fixing [Fix^-]), N levels, and a variety of control treatments. Although the experiments did not yield satisfactory results (Table 3), further experimentation along these lines should be pursued so that a better understanding of the mechanisms involved can be obtained. This will be a necessary prerequisite for the commercial implementation of this technology.

2. Enhanced N, P, and K Uptake

Although *Azospirillum* can use pectin as its sole carbon source, *A. brasilense* strain Sp 7 showed only slight pectinolytic (pectin lyase and endopolygalacturonase) activity (Tien et al., 1979; Umali-Garcia et al., 1980), as did extracts of inoculated wheat roots (Okon and Kapulnik, 1986). Such activity might be involved in a slight hydrolysis of the middle lamellae of *Azospirillum*-colonized cortical cells without causing cell collapse (appearance of electron transparent zones), which may accelerate water and nutrient uptake (Kapulnik et al., 1985a; Lin et al., 1983; Sarig et al., 1984; Umali-Garcia et al., 1980). Microscopic examination of roots of live maize and wheat plants shows distortions in the cortical cell arrangement caused by *Azospirillum*, indicating a weakening of natural adherence in cortical tissue of inoculated roots, which would increase mineral adsorption by a kind of "sponge" effect. Inoculation of maize with *A. brasilense* strain Cd or Sp 7 increased uptake of nitrate, potassium, and phosphate into excised root segments 30 to 50% over the controls, which was associated with altered cell arrangement in the outer four or five layers of the cortex. The enhanced growth of inoculated plants may be due to this increased uptake of nutrients (Okon and Kapulnik, 1986; Lin et al., 1983; Okon, 1982). Mechanisms other than N_2 fixation have been advanced to explain inoculation responses to *Azospirillum*, such as enhanced uptake of nitrate and P and K by roots of maize, wheat, and sorghum (Kapulnik et al., 1985a,c, 1987; Lin et al.,

1983; Okon, 1982), which varied with the strain, with *A. brasilense* strain Sp 245 increasing and strain Sp 7 decreasing P and K uptake (Jain and Patriquin, 1984). Murty and Ladha (1988) have also reported yield increases in rice inoculated with *A. lipoferum* strain 34-H associated with an increased root surface area and increased rates of NH_4 and PO_4 uptake. This enhanced nutritional status, which probably results from improved root surface area and activity, often leads to yield increases which, in wheat, are likely to stem from the increased number of fertile tillers per unit area in inoculated plants (Kapulnik et al. 1983, 1987; Millet et al., 1985; Reynders and Vlassak, 1982a,b). Sarig et al. (1988) have reported increased contents of N and P in the leaves of sorghum following inoculation with A. *brasilense,* which corresponded to the increases in biomass observed. Morgenstern and Okon (1987) showed that inoculation of *Sorghum bicolor* × *Sorghum sudanense* resulted in increased uptake of nitrate and uptake and accumulation of Rb, which was apparently due to the overall increase in root and shoot growth and not to a specific increase in nutrient uptake.

3. Plant Water Status

Okon et al. (1988b) have reported that the moisture status of maize and sorghum was favorably affected by *Azospirillum* inoculation, but they presented no supporting data. They claim that "significantly less pressure was needed to extract water from inoculated leaves than from controls and stomatal conductance was higher and canopy temperature lower in inoculated plants under water stress than in non-irrigated controls." In a subsequent paper, Sarig et al. (1988) demonstrated that inoculation resulted in higher leaf water potentials, lower canopy temperatures, and greater stomatal conductance and transpiration. Inoculated plants extracted more soil water, particularly from deeper layers, indicating that yield increases resulting from inoculation are primarily from improved soil moisture utilization.

B. Plant Growth Regulators

Azospirilla grown in culture are known to produce growth-promoting compounds, such as gibberellin-like and cytokinin like substances and auxins such as IAA from tryptophan (Tien et al., 1979). The general belief exists that *Azospirillum* increases root mass and function and changes root and root hair morphology (Brown, 1982; Inbal and Feldman, 1982; Jain and Patriquin, 1984; Kapulnik et al., 1981a, 1985a; Reynders and Vlassak, 1982b; Tien et al., 1979). Horemans and Vlassak (1985) demonstrated that *A. brasilense* could produce IAA in the absence of tryptophan when grown aerobically in the presence of NH_4, while De Francesco et al. (1985) showed that the highest levels of auxin were produced in both N_2-fixing conditions and limiting ammonia stationary cultures of *A. brasilense* strain Sp 6. Increased diameters of maize stems and increased tillering and number of ears in wheat as a result of inoculation may also be due to hormonal effects (Kapulnik et al., 1982; Reynders and Vlassak, 1982b; Zambre et al., 1984). Part of the evidence for the role of plant growth regulators in yield responses has been derived from the argument that if inoculation resulted in no increase in N concentration in the host tissue, then the increased growth would be due to these substances (Boddey and Dobereiner, 1988). This is a spurious argument because N concentration in tissue is a ratio (wt N/wt DM) × 100, in which both numerator and denominator can vary independently (Sumner, 1978). Thus the ratio would be maintained constant if the plant took up N at the same rate as it was accumulating dry matter. However, the reverse argument, namely, when a plant exhibits both increased N concentration in the tissue and increased yield (increased N uptake), would be evidence to show that some biological N_2 fixation may have taken place (Table 4). It is possible to mimic the effects of *Azospirillum* using a mixture of IAA, gibberellin GA-3, and kinetin, which increased root hairs and branching in *Pennisetum* (Table 5) (Tien et al., 1979). There was no synergistic effect of adding all three growth regulators together. As with bacterial concentrations, an optimal relationship was found between root elongation and the concentrations of IAA and cytokinin in barley, wheat, and *Panicum,* with no synergism from simultaneous addition (Okon and Kapulnik, 1986). Kolb and Martin (1985) have shown that spraying a solution of 10^{-9} g/L IAA on roots of wheat growing in root boxes resulted in a significant increase in root length, which mimicked *Azospirillum* inoculation. Spraying an inoculum of *A. brasilense* strain FT-326 on roots of *Beta vulgaris* resulted in significant increases in both root length and number of laterals. Another study showed that strains of *A. brasilense* that caused a maximum increase in plant growth in soil showed higher quantities of auxins in culture filtrates (Venkateswarlu and Rao, 1983). Barbieri et al. (1986) showed that plant response of *Triticum durum* in terms of increased number and length of lateral roots is a function of the ability to produce IAA but is not related to the N-fixing capacities of the *Azospirillum* strains (Table 6). The work of Jain and Patriquin (1985) strongly suggests that IAA is responsible for root hair branching in wheat, since it is affected by both plant and bacterial genomes. These genome effects can be explained by differen-

Table 5. Effect of plant growth hormones and inoculation with *A. brasilense* strain Sp 13t SR2 on the growth of pearl millet

Categories	Concentration (mg/L)	Root weight[a] (mg)	Top weight[a] (mg)
Control		161a	694b
IAA	0.01	200a	990a
GA-3	0.05	168a	832ab
Kinetin	0.001	181a	923a
IAA + GA + kinetin		155a	822ab
Inoculated		211a	1017a

From Tien et al. (1979).

[a]Values within columns followed by the same letter are not significantly different at the 0.05 level of probability.

Table 6. Effect of different strains of *A. brasilense* on root growth of Durum wheat and IAA production in culture

Strain	N fixation	IAA produced ($g/A_{670}/7$ days)	Lateral roots[a] Number	Lateral roots[a] Length (cm)
Uninoculated		0	8.25a	5.78a
Sp 6	++	7	15.67b	18.80b
Sp F103	– –	71	14.16b	22.67b
Sp F57	– –	0.4	4.20a	3.20a

From Barbieri et al. (1986).

[a]Values within columns followed by the same letter are not significantly different at the 0.05 level of probability.

tial plant sensitivity to IAA and differing abilities of bacteria to produce IAA. They suggested that this phenomenon is a predictor of the magnitude of plant growth response to inoculation. The role that root hair branching plays in root colonization by *Azospirillum* is not known, but the increased surface area resulting from branching is likely to be important in the acquisition of nutrients and water. Barbieri et al. (1988) showed that the number and length of lateral roots of *T. durum* were increased by inoculation with *A. brasilense* strains Sp 6 and Sp 7, but a non-IAA-producing mutant of strain Sp 7 had no effect. Total shoot and root weights were not affected by any inoculation. Fallik et al. (1989) showed that the roots of *Azospirillum*-inoculated maize seedlings contained elevated levels of both bound and free IAA and IBA (indole-3-butyric acid) two weeks after sowing. IBA is a commonly used commercial promoter of root initiation in cuttings. They suggested that the higher amounts of free IAA and IBA in inoculated roots may be associated with yield increases.

Although the effect of *Azospirillum* inoculation can be mimicked by applying growth hormones to roots, only circumstantial evidence exists to support claims that such hormonal activity takes place in situ in the rhizosphere. Because hormone levels in roots are very close to detection limits, the resolution of differences between treated and control plants is difficult. Therefore, most experiments have been conducted with additions of tryptophan from which IAA is formed to facilitate the detection of differences. Even the use of *A. brasilense* strain FT-326, which overproduces IAA, did not increase root length above wild selections, indicating that other hormones or factors may be involved as well. Furthermore, there is no direct evidence to support the claim that yield responses under field conditions to *Azospirillum* inoculation are the result of growth-promoting substances. However, Kapulnik et al. (1987) demonstrated enhanced root development and branching in the wheat cultivar Miriam under field conditions, which proved to be a decisive factor in permitting it to withstand droughty conditions and yield better than those cultivars that did not exhibit enhanced root growth. The possibility that roots produce plant growth-promoting substances in response to bacterial cell surface components or pectic enzymes of

Azospirillum needs to be investigated. Specific activities of polyphenoloxidase, peroxidase, IAA oxidase, and lactase were higher in extracts of *Azospirillum*-inoculated plants than in controls. This suggests that the former have a larger proportion of younger roots (Okon and Kapulnik, 1986). Fallik et al. (1988) found increased specific activities of the enzymes alcohol dehydrogenase, acid phosphatase, glutamine synthetase, isocitrate dehydrogenase, malate dehydrogenase, pyruvate kinase, and shikimate dehydrogenase in inoculated maize root extracts. As inoculation did not affect the activity of phenylalanine-ammonia-lyase and glucose-6-phosphate dehydrogenase, which are usually higher in plants infected with microorganisms, it would appear that *A. brasilense* does not behave as a pathogenic rhizosphere organism. While evidence has accumulated in favor of responses being at least partly due to growth-promoting substances on inoculation with *Azospirillum*, the origin of these growth factors, microbial or plant, has not yet been clarified. There is recent evidence to suggest that plant-produced hormones such as IAA, 6-benzyl-aminopurine, and gibberellin may directly affect the N_2-fixing ability of *Azospirillum* (Christiansen-Weniger, 1988).

Okon et al. (1988a) have proposed a hypothetical model to explain responses to *Azospirillum* inoculation. Once roots emerge from the seed glycolysis, the tricarboxylic acid (TCA) cycle and conversion of conjugate IAA to the active form are stimulated, which causes roots to assimilate more water and nutrients, leading to greater shoot growth which, in turn, supplies more photosynthate to roots. Under appropriate agronomic conditions, it is proposed that these processes will lead to increased yields. This hypothesis requires confirmation.

V. Importance of Strain Selection

Yield responses to *Azospirillum* are often unpredictable, which may in part be due to affinities but not necessarily absolute specificities of certain strains for their original host or group of hosts. The characteristics of many of the strains commonly used are presented in Table 7, from which some idea of likely affinities can be gleaned. The requirement of "homologous" strains (isolated from surface-sterilized roots of the same crop) for successful inoculation has been well established, with nonhomologous strains proving to be much less effective (Baldani and Dobereiner, 1980; Baldani et al., 1986, 1987; Boddey et al. 1986; Dobereiner and Baldani, 1981; Jain and Patriquin, 1984; Patriquin et al., 1983; Reynders and Vlassak, 1982a, 1982b). Despite the fact that this knowledge has been available for some time, many experiments are still conducted with no reference to the selection of appropriate inocula (Raverkar and Konde, 1988; Schmidt et al., 1988; Tilak and Subba Rao, 1987). With wheat, Reynders and Vlassak (1982a,b) increased yield more with *A. brasilense* strain Sp BR14 isolated from wheat than with strain S-631 isolated from maize (Table 4). In a field experiment with wheat in which several strains of *Azospirillum* isolated from wheat and maize were used, those strains isolated from surface-

Table 7. Characteristics and origin of *Azospirillum* spp. strains

Strain	Characteristics	Origin
Sp 245	*A. brasilense* nir^-, resistant to 20 $\mu g/mL$ spectinomycin	Chloramine-*t* treated wheat roots from monoculture field, Parana, Brazil
Sp 245 spec	Spontaneous mutant of Sp 245, resistant to 60 $\mu g/mL$ spectinomycin	
Sp 245 NR^- (Sp 245 spec nr^-)	Spontaneous mutant, selected in nutrient agar containing $NaClO_3^-$ and NO_2^- as *e* acceptor	Chloramine-*t* treated wheat roots from monoculture field, Parana, Brazil
Sp 246 spec	*A. brasilense,* resistant to 60 $\mu g/mL$ spectinomycin	Washed wheat roots grown in field, Parana, Brazil
Sp 107	*A. brasilense* nir^-	Chloramine-*t* treated wheat roots grown in pots of Itaguai soil, Rio de Janeiro
Sp 107st	*A. brasilense* nir^-, resistant to 150 $\mu g/mL$ streptomycin	Chloramine-*t* treated wheat roots grown in pots, Rio de Janeiro
Tr1	*A. brasilense,* resistant to 60 $\mu g/mL$ streptomycin	Washed wheat roots from field, Parana, Barsil
S-7	*A. brasilense* nir^+	Isolated from *Eragrostis tremula* surface-sterilized roots grown in Indian desert soil
S-14	*A. brasilense* nir^+	Isolated from *Cynodon dactylon* surface-sterilized roots grown in Indian desert soil
S-51	*A. brasilense*	Isolated from *Cenchrus cilaris* surface-sterilized roots grown in Indian desert soil
S-54	*A. brasilense* nir^+	Isolated from *Pennisetum typhoides* surface-sterilized roots grown in Indian desert soil
S-59	*A. brasilense*	Isolated from *Lasiurus sindicus* surface-sterilized roots grown in Indian desert soil
JM 6A2	*A. brasilense* nir^+	Isolated from maize roots, Ecuador
Sp7	*A. brasilense* nir^+ (ATCC 29145)	Isolated from *Digitaria* rhizoshpere field soil, Rio de Janeiro
Sp 7 nad st (Sp FP2)	*A. brasilense* nir^+ (ATCC 29145), resistant to 15 $\mu g/mL$ nalidixic acid and 200 $\mu g/mL$ streptomycin	Isolated from *Digitaria* rhizosphere field soil, Rio de Janeiro
Sp 6	Same as Sp 7	

Table 7. *Continued*

Strain	Characteristics	Origin
Sp F104	*A. brasilense* fix$^-$, non N fixing	Isolated after nitrosoguanidine mutagenesis from Sp 6
Sp F105	Similar to Sp F104	
Cd	*A. brasilense* (ATCC 29729), red pigment	Isolated from pots with *Cynodon* inoculated with Sp 7
CdSR	*A. brasilense* (ATCC 29729), red pigment, streptomycin and rifampicin resistant	
Cd-1	*A. brasilense*	Isolated from *Cynodon dactylon* grown in sandy Israeli soil
S-631	*A. brasilense*	Isolated from sandy loam maize soil, Belgium
Sp Br14	*A. brasilense* nir$^-$	Isolated from wheat rhizosphere soil, Brazil
E-15	*A. brasilense*	Isolated from maize roots in Egyptian soil
M-4	*A. brasilense*	Isolated from maize root surface in Indonesian soil
SE	*A. brasilense*	Isolated from clay soil cultivated to wheat, Egypt
SC-22	*A. brasilense*	Isolated from clay soil cultivated to clover, Egypt
Sp 262 chl kan	*A. brasilense*, resistant to 25 μg/mL chloramphenicol and 10 μg/mL kanamycin	Chloramine-*t* treated *Setaria* roots grown in pots in soil–vermiculite mixture, Rio de Janeiro
A-95	*A. brasilense*	Isolated from rice rhizosphere soil, France
RO7	*A. brasilense*	Isolated from rice rhizosphere soil, Senegal
FT-326	*A. brasilense* (overproduces IAA)	Isolated from *Beta vulgaris*
Sp S82 chl st	*A. lipoferum*, resistant to 10 μg/mL chloramphenicol and 100 μg/mL streptomycin	Chloramine-*t* treated sorghum roots grown in field, Rio de Janeiro
S-82	*A. lipoferum*	Chloramine-*t* treated sorghum roots
S-19	*A. lipoferum*	Chloramine-*t* treated sorghum roots
S-65	*A. lipoferum*	Washed sorghum roots
ICM 1001	*A. lipoferum*	Isolated from sorghum (CSH 1) rhizosphere in vermiculite inoculated with Napier bajra root extract at ICRISAT

Table 7. *Continued*

Strain	Characteristics	Origin
34-H	*A. lipoferum*	Isolated from rice roots
4B	*A. lipoferum*	Isolated from rice rhizosphere soil, France
4T	*A. lipoferum* (nonmotile)	Isolated from rice rhizosphere soil, France
B7C	*A. lipoferum*	Isolated from maize rhizosphere soil, France
Sp 108st	*Azospirillum* sp. nir^-, resistant to 150 μg/mL streptomycin	Chloramine-*t* treated maize roots grown in pots of red yellow podzolic soil, Rio de Janeiro
Sp 242st	*Azospirillum* sp. nir^-, resistant to 150 μg/mL streptomycin	Chloramine-*t* treated maize roots grown in field in hydromorphic soil, Rio de Janeiro
YTr1	*A. amazonense*, resistant to 20 μg/mL streptomycin	Washed wheat roots from field, Rio de Janeiro
Am Ytr st	*A. amazonense*, resistant to 20 μg/mL streptomycin	Washed wheat roots grown in field, Rio de Janeiro

sterilized wheat roots (Sp 245, Sp 107) were consistently better in colonizing the root interior than those from wheat rhizosphere soil (Sp Br14) or surface-sterilized maize roots (Sp 242, Sp 108). Only strains Sp 245 and Sp 107 resulted in yield and N uptake increases (Table 4), with a highly significant correlation ($r = 0.92$) between the number of *Azospirillum* within the root and total N accumulation but no relationship with the number of bacteria from nonsterile (washed) roots (Baldani et al., 1983). This evidence strongly supports the view that the responses to inoculation are due to those bacteria that penetrate the root. Recently Ferreira et al. (1987) clearly demonstrated the superiority of *A. brasilense* strain Sp 245 for the inoculation of wheat, confirming similar previous observations in Brazil, which showed that "homologous" strains are consistently able to establish themselves in the interior of roots of wheat and sorghum (Baldani et al., 1986, 1987). Where positive yield responses have been obtained, microbiological investigations have always shown that the strain has been well established within the root. Jain and Patriquin (1984) found that *A. brasilense* strain Sp 7 (isolated from *Digitaria*) reduced the root mass of wheat. Greater responses in dry matter and grain yield were obtained with pearl millet and wheat when homologous rather than nonhomologous strains were used (Reynders and Vlassak, 1982a,b; Wani et al., 1985). Christiansen-Weniger et al. (1985) demonstrated that inoculations with *A. lipoferum* strains S-82 and S-19 isolated from within the root of sorghum were much more effective than those from the root exterior (S-65) or rhizosphere soil (*A. brasilense* strain Sp 7). Bouton et al. (1985) demonstrated a high degree of specificity between inbred lines of pearl millet and *A. brasilense* strain JM 125 when assayed as ARA. This finding points to the importance of plant breeding in enhancing this association and thereby producing germplasms useful in agronomic and physiological evaluations of associative N_2 fixation. In a study involving a number of strains, O'Hara et al. (1987) found that maize responded to *A. brasilense* strain Sp 7 and pearl millet to strain Sp 107st, while strain Sp F105 caused a reduction in pearl millet. In two separate experiments, maize responded to strain Sp 107st in one but not in the other. These observations are not quite in line with the homologous strain hypothesis of the Brazilian school, but this work was carried out in sterile sand–solution culture where there would be no competition for inoculation sites. In Israel, substantial dry matter and N-yield responses (Table 4) in a number of cereal crops have been obtained with nonhomologous *A. brasilense* strains Sp 7 and Cd (Kapulnik et al., 1981c, 1987; Sarig et al., 1984; Yahalom et al., 1984), but it is the view of Boddey and Dobereiner (1988) that these responses are due to the lack of competition in Israeli soils, which have very low to zero native *Azospirillum* populations, allowing virtually any strain to establish and colonize roots. This view is supported by the results of Boddey et al. (1986), who obtained total grain N-yield responses in sterilized soil irrespective of whether homologous or nonhomologous strains were used (Table 4). Further evidence of the specificity of *Azospirillum* strains was cited by Baldani et al. (1986) in relation to chemotaxis to various organic acids and to root tip exudates in relation to attachment to roots. Responses to inoculation appear to depend on both the bacterial and the host genomes at the cultivar and subspecies levels as well as on a multitude of other factors, all of which contribute to the unpredictability observed.

VI. Conclusions

Azospirillum brasilense and *A. lipoferum* form associations with C-4 grasses and C-3 grasses and sugarcane, respectively, which are mutually beneficial to both the host plant and the bacterium. There is also evidence of beneficial associations with dicotyledonous plants, especially legumes.

Initially the bacteria are adsorbed on the root surfaces with greater or lesser degrees of specificity. There is strong evidence to support the existence of a recognition process similar to that in operation between pathogens and their specific hosts. This specificity is of great importance to the beneficial association between organism and higher plant without which little benefit is likely to accrue and is determined at the genome level of both host and bacterium. A summary of the possible mechanisms by which *Azospirillum* inoculation results in improved crop growth has been presented by Okon et al. (1988) in Table 8.

In cases where N_2 fixation plays a substantial role in the growth response obtained, it would be necessary for the bacterium to locate at a position where it would be able to control the supply of oxygen. Internal root colonization by *Azospirillum*, which has been demonstrated in a number of cases, would go a

Table 8. Possible mechanisms of action of *Azospirillum* inoculation

A.	Effects on roots are caused by colonization of a critical number of sites on the root by a critical number of *Azospirillum* cells.
B.	Contribution of biological nitrogen fixation
	1. Acetylene reduction activity at flowering
	2. Very small amounts of N incorporated in plants ($^{15}N_2$ fixation)
	3. Possible enrichment of soil with combined nitrogen
	4. Important for long-term N balance in ecosystem
	5. Apparently of no major significance in modern agriculture
C.	Production of plant growth-promoting substances (bacteria)
	1. Production by *Azospirillum* in culture
	2. Not demonstrated to take place in association with plants
	3. May increase development of root hairs and branching of roots
D.	Production of plant growth-promoting substances (plant)
	1. *Azospirillum* surface components or pectic enzymes elicit plant responses, production of plant growth-promoting substances by plant (imbalance?)
	2. Results in morphological changes in root cells
	3. Results in decreased activity of oxidative enzymes (IAA oxidase)
	4. Increase in endogenous IAA
	5. Results in increased mineral and water uptake, root development, vegetative growth, and crop yield
E.	Competition in the rhizosphere with other damaging microorganisms
	1. Antagonism
	2. Displacement
	3. Competition for nutrients

From Okon et al., Gordon and Breach Science Publishers, 1988.

long way to achieving this end. The formation of protective membranes, which has been observed, may also serve this purpose.

The beneficial effect on the higher plant of this association is likely to result from a combination of N_2 fixation and the growth-promoting effects of plant hormones by the bacterium that usually stimulates root growth, branching, and surface area. Despite the fact that *Azospirillum* spp. are known to fix N_2, the quantity supplied to the plant is only of minor significance and nowhere nearly meets its N requirements. As a result of this improved rooting habit arising from inoculation, plants often take up more nutrients, including N, usually as nitrate. There appears to be a plethora of *Azospirillum* strains that have been isolated and tested by various workers. With very few exceptions, they have seldom been tested under similar conditions, with the result that it is very difficult to determine which strains are most effective. In terms of selection and breeding programs for both host and bacterium, it seems that the best approach would be one of selecting for improved vigor and yield rather than for N_2 fixation in the association.

As a result of the combination of improved N nutrition from fixation by the bacterium, increased root surface area, and improved nutrient and water uptake, plants often show yield responses to inoculation with appropriately matched azospirilla. The magnitude of these responses has often been quite substantial, well in excess of the likely costs of inoculation, which would make this technique highly attractive to the farmer. Unfortunately this yield-promoting effect is not always observed, making inoculation a somewhat unpredictable treatment. The reasons for this unpredictability can range from inappropriately matched inoculum to external conditions preventing the expression of the effect, such as severe drought or other stresses. The most recent results in the literature would suggest that special attention should be paid to making sure that homologous strains of *Azospirillum* are used for inoculation as well as in sufficient concentrations to outcompete native less effective strains. Despite the fact that this has been known for some time, there are still many reports where the nature of the inoculating organism is very poorly described, which inevitably leads to confusion and may contribute to the lack of reproducibility in published results. In addition, more attention should be paid to promoting the survival of *Azospirillum* cells between the time they are applied to the soil and the moment of colonization of the root. It may be possible, by breeding and genetic manipulation, to increase the degree of specificity between bacterium and host, thus ensuring that the organisms in the inoculum dominate in the rhizosphere and within the root.

In reviewing the literature, one is often faced with the dilemna of knowing whether there may not have been many negative or zero responses to inoculation that were never written up. In a number of cases during the preparation of this review, situations have been identified where authors reporting on the work of others have engaged in selective presentation of results, which causes distortions of the picture in the mind of the reader. This is a particularly undesirable state of affairs, which should be avoided.

To achieve success with this technology, it will be necessary to conduct field experiments comparing various homologous strains of *Azospirillum* at appropriate rates and times of inoculation on a number of crops under a variety of environmental conditions and seasons to ascertain which ones are most effective in promoting crop yield. Unfortunately surprisingly few such studies have been conducted. After it has been demonstrated that consistent yield responses to inoculation with appropriate strains can be obtained, it will be necessary to develop suitable carriers for the bacteria to ensure long shelf life and effectiveness. Given additional research in this and other areas, it should be possible, in the foreseeable future, to develop this technology to the stage where it can be exploited commercially. There appear to be very few opportunities for the exclusive marketing of *Azospirillum* strains because they can be easily cultured.

References

Albrecht, S. L., Y. Okon, J. Lonnquist, and R. H. Burris. 1981. Nitrogen fixation by corn *Azospirillum* associations in a temperate climate. *Crop Sci.* 21:301–306.

Albrecht, S. L., M. H. Gaskins, J. R. Milam, S. C. Schank, and R. L. Smith. 1983. Ecological factors affecting survival and activity of *Azospirillum* in the rhizosphere. In: W. Klingmuller (ed.), *Azospirillum II*. Birkhauser, Basel.

Avivi, Y., and M. Feldman. 1982. The response of wheat to bacteria of the genus *Azospirillum*. *Israel J. Bot.* 31:237–245.

Baldani, V.L.D., and J. Dobereiner. 1980. Host plant specificity in the infection of cereals with *Azospirillum* spp. *Soil Biol. Biochem.* 12:433–439.

Baldani, V.L.D., J. I. Baldani, and J. Dobereiner. 1983. Effect of *Azospirillum* inoculation on root infection and nitrogen incorporation in wheat. *Canad. J. Microbiol.* 29:924–929.

Baldani, V.L.D., M. A. Alvarez, J. I. Baldani, and J. Dobereiner. 1986. Establishment of inoculated *Azospirillum* spp. in the rhizosphere and in roots of field grown wheat and sorghum. *Plant Soil* 90:35–46.

Baldani, V.L.D., J. I. Baldani, and J. Dobereiner. 1987. Inoculation of field grown wheat *(Triticum aestivum)* with *Azospirillum* spp. in Brazil. *Biol. Fertil. Soils.* 4:37–40.

Baltensperger, A. A., S. C. Schank, R. L. Smith, R. C. Littell, J. H. Bouton, and A. E. Dudeck. 1978. Effect of inoculation with *Azospirillum* and *Azotobacter* on turf-type bermuda genotypes. *Crop Sci.* 18:1043–1045.

Barbieri, P., T. Zanelli, E. Galli, and G. Zanetti. 1986. Wheat inoculation with *Azospirillum brasilense* Sp 6 and some mutants altered in nitrogen fixation and indole-3-acetic acid production. *FEMS Microbiol. Lett.* 36:87–90.

Barbieri, P., A. Bernardi, E. Galli, and G. Zanetti. 1988. Effects of inoculation with different strains of *Azospirillum brasilense* on wheat root development. In: W. Klingmuller (ed.), *Azospirillum IV: genetics, physiology, ecology*. Springer-Verlag, Berlin.

Bashan, Y., and H. Levanony. 1988. Interaction between *Azospirillum brasilense* CD and wheat root cells during early stages of root colonization. In: W. Klingmuller (ed.), *Azospirillum IV: genetics, physiology, ecology*. Springer-Verlag, Berlin.

Boddey, R. M., and J. Dobereiner. 1982. Association of *Azospirillum* and other diazotrophs with tropical gramineae. In: *Non-Symbiotic Nitrogen Fixation and Organic Matter in the Tropics. Trans. 12th Int. Cong. Soil Sci.*, New Dehli.

Boddey, R. M., and J. Dobereiner. 1988. Nitrogen fixation associated with grasses and cereals: Recent results and perspectives for future research. *Plant Soil* 108:53–65.

Boddey, R. M., V. L. Baldani, J. I. Baldani, and J. Dobereiner. 1986. Effect of inoculation of *Azospirillum* spp. on nitrogen accumulation by field grown wheat. *Plant Soil* 95:109–121.

Bouton, J. H. 1988. Acetylene reduction activity of pearl millet inbred lines grown in soil. *Plant Soil* 110:143–144.

Bouton, J. H., and D. A. Zuberer. 1979. Response of *Panicum maximum* to inoculation with *Azospirillum brasilense*. *Plant Soil* 52:585–590.

Bouton, J. H., R. L. Smith, S. C. Schank, G. W. Burton, M. E. Tyler, R. C. Littel, R. N. Gallaher, and K. H. Quesenberry. 1979. Response of pearl millet inbreds and hybrids to inoculation with *Azospirillum brasilense*. *Crop Sci.* 19:12–16.

Bouton, J. H., S. L. Albrecht, and D. A. Zuberer. 1985. Screening and selection of pearl millet for root associated bacterial nitrogen fixation. *Field Crops Res.* 11:131–140.

Boyle, C. D., and D. G. Patriquin. 1980. Endorhizal and exorhizal acetylene reducing activity in a grass (*Spartina alterniflora* Loisel.)—diazotroph association. *Plant Physiol.* 66:276–280.

Brown, M. E. 1982. Nitrogen fixation by free-living bacteria associated with plants—fact or fiction? In: M. Rhodes-Roberts and F. A. Skinner (eds.), *Bacteria and plants*. Academic Press, London.

Charyulu, P.B.B.N., F. Fourcassie, A. K. Barbouche, L. Rondro Harisoa, A. M. N. Omar, P. Weinhard, R. Marie, and J. Balandreau. 1985. Field inoculation of rice using in vitro selected bacterial and plant genotypes. In: W. Klingmuller (ed.), *Azospirillum III: genetics, physiology, ecology*. Springer-Verlag, Berlin.

Christiansen-Weniger, C. 1988. An influence of plant growth substances on growth and nitrogenase activity from *Azospirillum brasilense*. In: W. Klingmuller (ed.), *Azospirillum IV: genetics, physiology, ecology*. Springer-Verlag, Berlin.

Christiansen-Weniger, C., R. M. Boddey, and J. Dobereiner. 1985. Evaluation of nitrogen fixation in sorghum cultivars innoculated with different strains of *Azospirillum* spp. In: W. Klingmuller (ed.), *Azospirillum III: genetics, physiology, ecology*. Springer-Verlag, Berlin.

Cohen, E., Y. Okon, J. Kigel, I. Nur, and Y. Henis. 1980. Increase in dry weight and total nitrogen content in *Zea mays* and *Setaria italica* associated with nitrogen fixing *Azospirillum* spp. *Plant Physiol.* 66:746–749.

Danneberg, G., W. Zimmer, and H. Botle. 1985. Some physiological and biochemical properties of denitrification by *Azospirillum brasilense*. In: W. Klingmuller (ed.), *Azosperillum III: genetics, physiology, ecology*. Springer-Verlag, Berlin.

Darmwal, N. S., and A. S. Gaur. 1988. Associative effect of celluloytic fungi and *Azospirillum lipoferum* on yield and nitrogen uptake by wheat. *Plant Soil* 107:211–218.

De Francesco, R., G. Zanetti, P. Barbieri, and E. Galli. 1985. Auxin production by *Azospirillum brasilense* under different cultural conditions. In: W. Klingmuller (ed.), *Azospirillum III: genetics, physiology, ecology*. Springer-Verlag, Berlin.

Dobereiner, J., and V.L.D. Baldani. 1981. Prospects for innoculation of grasses with *Azospirillum* spp. pp. 1–9. In: P. B. Vose and A. P. Ruschel (eds.), *Associative N_2 Fixation*, vol. 2. CRC Press, Boca Raton, FL.

Dobereiner, J., and J. M. Day. 1976. Associative symbioses in tropical grasses: characterization of microorganisms and dinitrogen-fixing sites. In: W. E. Newton and C. J. Nyman (eds.), *Proc. 1st Int. Symp. on N_2 Fixation*. Washington Univ. Press, Pullman, WA.

Eyers, M., F. Waelkens, J. Vanderleyden, and A. P. Van Gool. 1988. Quantitative measurement of *Azospirillum* plant cell attachment. In: W. Klingmuller (ed.), *Azospirillum IV: genetics, physiology, ecology*. Springer-Verlag, Berlin.

Fallik, E., Y. Okon, and M. Fischer. 1988. The effect of *Azospirillum brasilense* inoculation on metabolic enzyme activity in maize root seedlings. *Symbiosis* 6:17–28.

Fallik, E., Y. Okon, E. Epstein, A. Goldman, and M. Fischer. 1989. Identification and quantification of IAA and IBA in *Azospirillum brasilense*-inoculated maize roots. *Soil Biol. Biochem.* 21:147–153.

Fayez, M., and Z. Y. Daw. 1987. Effect of inoculation with different strains of *Azospirillum brasilense* on cotton *(Gossipium barbadense)*. *Biol. Fertil. Soils* 4:91–95.

Fayez, M., N. F. Emam, and H. E. Makboul. 1985. The possible use of nitrogen fixing *Azospirillum* as biofertilizer for wheat plants. *Egypt. J. Microbiol.* 20:199–206.

Ferreira, M.C.B., M. S. Fernandes, and J. Dobereiner. 1987. Role of *Azospirillum brasilense* nitrate reductase in nitrate assimilation by wheat plants. *Biol. Fertil. Soils* 4:47–53.

Gadkari, D. 1988. Influence of herbicides on growth and nitrogenase activity of *Azospirillum*. In: W. Klingmuller (ed.), *Azospirillum IV: genetics, physiology, ecology*. Springer-Verlag, Berlin.

Haahtela, K., T. Laasko, E. Nurmiaho-Lassila, and T. K. Korhonen. 1988. Effects of inoculation of *Poa pratensis* and *Triticum aestivum* with root-associated N_2 fixing *Klebsiella, Enterobacter* and *Azospirillum*. *Plant Soil* 106:239–248.

Hadas, R., and Y. Okon. 1988. Effect of *Azospirillum brasilense* inoculation on root morphology and respiration in tomato seedlings. *Biol. Fertil. Soils* 5:241–247.

Harris, J. M., J. A. Lucas, M. R. Davey, G. Lethbridge, and K. A. Powell. 1989. Establishment of *Azospirillum* inoculant in the rhizosphere of winter wheat. *Soil Biol. Biochem.* 21:59–64.

Hegazi, N. A. 1982. Contribution of *Azospirillum* spp. to asymbiotic N_2-fixation in soils and on roots of plants grown in Egypt.

Hegazi, N. A., and H. Saleh. 1985. Possible contribution of *Azospirillum* spp. to the nutritional status of wheat plants grown in sandy soils of Gassim-Saudi Arabia. In: W. Klingmuller (ed.), *Azospirillum III: genetics, physiology, ecology*. Springer-Verlag, Berlin.

Hegazi, N. A., M. Monib, and K. Vlassak. 1979. Effect of inoculation with N_2 fixing spirilla and *Azotobacter* on nitrogenase activity on roots of maize grown under subtropical conditions. *Appl. Envir. Microbiol.* 38:621–625.

Hegazi, N. A., M. Monib, H. A. Amer, and E. Shokr. 1983. Response of maize plants to inoculation with *Azospirilla* and (or) straw amendment in Egypt. *Canad. J. Microbiol.* 29:888–894.

Hess, D., D. Heinrich, J. Kemmner, and S. Jekel. 1985. Secretion and cleavage of sucrose by wheat: A system of chemotactic attraction and nitrogenase induction in *Azospirillum lipoferum*. In: W. Klingmuller (ed.), *Azospirillum III: genetics, physiology, ecology*. Springer-Verlag, Berlin.

Heulin, T., A. Guckert, and J. Balandreau. 1987. Stimulation of root exudation of rice seedlings by *Azospirillum* strains: carbon budget under gnotobiotic conditions. *Biol. Fertil. Soils* 4:9–14.

Horemans, S., and K. Vlassak. 1985. Production of indol-3-acetic acid by *Azospirillum brasilense*. In: W. Klingmuller (ed.), *Azospirillum III: genetics, physiology, ecology*. Springer-Verlag, Berlin.

Horemans, S., K. De Coninck, and K. Vlassak. 1988. Aspects of the ecology of *Azospirillum* sp. in Belgian soils. In: W. Klingmuller (ed.), *Azospirillum IV: genetics, physiology, ecology*. Springer-Verlag, Berlin.

Hubbell, D. H. 1981. Legume infection by *Rhizobium:* a conceptual approach. *Bioscience* 31:832–836.

Hubbell, D. H., and M. H. Gaskins. 1984. Associative N_2 fixation with *Azospirillum*. In: M. Alexander (ed.), *Biological Nitrogen Fixation*. Plenum, NY.

Inbal, E., and M. Feldman. 1982. The response of a hormonal mutant of common wheat to bacteria of the genus *Azospirillum*. *Israel J. Bot.* 31:257–263.

Ishac, Y. Z., M. E. El-Haddad, M. J. Daft, E. M. Ramadan, and M. E. El-Demerdash. 1986. Effect of seed inoculation, mycorrhizal infection and organic amendment on wheat growth. *Plant Soil* 90:373–382.

Jagnow, G. 1985. Root colonization and plant growth of grasses and cereals in unsterile soil and nitrogenase (C_2H_2) activity of sterile cereal seedlings after inoculation with *Azospirillum* spp. In: W. Klingmuller (ed.), *Azospirillum III: genetics, physiology, ecology*. Springer-Verlag, Berlin.

Jain, D. K., and D. G. Patriquin. 1984. Root hair deformation, bacterial attachment and plant growth in wheat *Azospirillum* associations. *Appl. Envir. Microbiol.* 48:1208–1213.

Jain, D. K., and D. G. Patriquin. 1985. Characterization of a substance produced by *Azospirillum* which causes branching of wheat root hairs. *Canad. J. Microbiol.* 31:206–210.

Kannaiyan, S., M. Thangaraju, and G. Oblisami. 1983. Rice crop response to biofertilizers inoculation. *Madras Agric. J.* 70:65–66.

Kapulnik, Y., and Y. Okon. 1983. Benefits of *Azospirillum* spp. inoculation on wheat:

Effects on root development, mineral uptake, nitrogen fixation and crop yield. In: W. Klingmuller (ed.), *Azospirillum II*. Birkhauser, Basel.

Kapulnik, Y., J. Kigel, Y. Okon, I. Nur, and Y. Henis. 1981a. Effect of *Azospirillum* inoculation on some growth parameters and N content of wheat, sorghum and panicum. *Plant Soil* 61:65–70.

Kapulnik, Y., Y. Okon, J. Kigel, I. Nur, and Y. Henis. 1981b. Effects of temperature, nitrogen fertilization and plant age on nitrogen fixation by *Setaria italica* inoculated with *Azospirillum brasilense* (strain cd). *Plant Physiol.* 68:340–343.

Kapulnik, Y., S. Sarig, I. Nur, Y. Okon, J. Kigel, and Y. Henis. 1981c. Yield increases in summer cereal crops in Israeli fields inoculated with *Azospirillum*. *Expl. Agric.* 17:179–187.

Kapulnik, Y., S. Sarig, I. Nur, Y. Okon, and Y. Henis. 1982. The effect of *Azospirillum* inoculation on growth and yield of corn. *Israel J. Bot.* 31:246–255.

Kapulnik, Y., S. Sarig, I. Nur, and Y. Okon. 1983. Effect of *Azospirillum* inoculation on yield of field grown wheat. *Canad. J. Microbiol.* 29:895–899.

Kapulnik, Y., Y. Okon, and Y. Henis. 1985a. Changes in root morphology of wheat caused by *Azospirillum* inoculation. *Canad. J. Microbiol.* 31:881–887.

Kapulnik, Y., M. Feldman, Y. Okon, and Y. Henis. 1985b. Contribution of nitrogen fixed by *Azospirillum* to the N nutrition of spring wheat in Israel. *Soil Biol. Biochem.* 17:509–515.

Kapulnik, Y., R. Gafny, and Y. Okon. 1985c. Effect of *Azospirillum* spp. inoculation on root development and nitrate uptake in wheat (*Triticum aestivum* cv Miriam) in hydroponic systems. *Canad. J. Bot.* 63:627–631.

Kapulnik, Y., Y. Okon, and Y. Henis. 1987. Yield response of spring wheat cultivars (*Triticum aestivum* and *T. turgidum*) to inoculation with *Azospirillum brasilense* under field conditions. *Biol. Fertil. Soils* 4:27–35.

Kolb, W., and P. Martin. 1985. Response of plants to inoculation with *Azospirillum brasilense* and to application of indole acetic acid. In W. Klingmuller (ed.), *Azospirillum III: genetics, physiology, ecology*. Springer-Verlag, Berlin

Lethbridge, G., and M. S. Davidson. 1983. Root-associated nitrogen-fixing bacteria and their role in the nitrogen nutrition of wheat estimated by ^{15}N isotope dilution. *Soil Biol. Biochem.* 15:365–374.

Lethbridge, G., M. S. Davidson, and G. P. Sparling. 1982. Critical evaluation of the acetylene reduction test for estimating the activity of nitrogen-fixing bacteria associated with the roots of wheat and barley. *Soil Biol. Biochem.* 14:27–35.

Lima, E., R. M. Boddey, and J. Dobereiner. 1987. Quantification of biological nitrogen fixation associated with sugarcane using a ^{15}N aided nitrogen balance. *Soil Biol. Biochem.* 19:165–170.

Lin, W., Y. Okon, and R.W.F. Hardy. 1983. Enhanced mineral uptake by *Zea mays* and *Sorghum bicolor* roots inoculated with *Azospirillum brasilense*. *Appl. Envir. Microbiol.* 45:1775–1779.

Malik, K. A., Y. Zafar, R. Bilal, and F. Azam. 1987. Use of ^{15}N isotope dilution for quantification of N_2 fixation associated with roots of kallar grass *(Leptochloa fusca)*. *Biol. Fertil. Soils* 4:103–108.

Mano, D.M.S., A.C.M. Matos and T. Langenbach. 1988. The effect of Difocol on morphology, growth and nitrogenase activity of *Azospirillum lipoferum*. In: W. Klingmuller (ed.), *Azospirillum IV: genetics, physiology, ecology*. Springer-Verlag, Berlin.

Marocco, A., and C. Lorenzoni. 1984. Cereal and grass species response to *Azospirillum brasilense* inoculation. *Ital. Soc. Agric. Genetics* (Abstracts).

Mertens, T., and D. Hess. 1984. Yield increases in spring wheat *(Triticum aestivum)* inoculated with *Azospirillum lipoferum* under greenhouse and field conditions of a temperate region. *Plant Soil* 82:87–99.

Millet, E., Y. Avivi, and M. Feldman. 1985. Effect of rhizosphere bacteria on wheat yield under field conditions. *Plant Soil* 86:347–355.

Morgenstern, E., and Y. Okon. 1987. Promotion of plant growth and NO_3 and Rb uptake in *Sorghum bicolor* × *Sorghum sudanense* inoculated with *Azospirillum brasilense*-Cd. *Arid Soil Res. Rehab.* 1:211–217.

Murty, M. G., and J. K. Ladha. 1987. Differential colonization of *Azospirillum lipoferum* on roots of two varieties of rice (*Oryza sativa* L.) *Biol. Fertil. Soils* 4:3–7.

Murty, M. G., and J. K. Ladha. 1988. Influence of *Azospirillum* inoculation on the mineral uptake and growth of rice under hydroponic conditions. *Plant Soil* 108:281–285.

Nayak, D. N., J. K. Ladha, and I. Wanatabe. 1986. The fate of marker *Azospirillum lipoferum* inoculated into rice and its effect on growth, yield and N_2 fixation of plants studied by acetlyene reduction, $^{15}N_2$ feeding and ^{15}N dilution techniques. *Biol. Fertil. Soils* 2:7–14.

Neuer, G., A. Kronenberg, and H. Bothe. 1985. Denitrification and nitrogen fixation by *Azospirillum* III. Properties of a wheat *Azospirillum* association. *Arch. Microbiol.* 139:364–370.

Nur, I., Y. Okon, and Y. Henis. 1980. An increase in nitrogen content of *Setaria italica* and *Zea mays* inoculated with *Azospirillum. Canad. J. Microbiol.* 26:482–485.

O'Hara, G. W., M. R. Davey, and J. A. Lucas. 1981. Effect of inoculation of *Zea mays* with *Azospirillum brasilense* strains under temperate conditions. *Canad. J. Microbiol.* 27:871–877.

O'Hara, G. W., M. R. Davey, and J. A. Lucas. 1987. Effect of nitrogen on the yield response of *Pennisetum americanum, Triticum aestivum* and *Zea mays* to inoculation with *Azospirillum brasilense* under temperate conditions. *Biol. Fertil. Soils* 4:67–72.

Okon, Y. 1982. *Azospirillum:* Physiological properties, mode of association with roots and its application for the benefit of cereal and forage crops. *Israel J. Bot.* 31:214–220.

Okon, Y. 1985. *Azospirillum* as a potential inoculant for agriculture. *Trends Biotech.* 3:223–228.

Okon, Y., and Y. Kapulnik. 1986. Development and function of *Azospirillum*-inoculated roots. *Plant Soil* 90:3–16.

Okon, Y., P. G. Heytler, and R.W.F. Hardy. 1983. N_2 fixation by *Azospirillum brasilense* and its incorporation into host *Setaria italica. Appl. Envir. Microbiol.* 46:694–697.

Okon, Y., E. Fallik, S. Sarig, E. Yahalom, and S. Tal. 1988a. Plant growth promoting effects of *Azospirillum.* In: H. Bothe, F. J. de Bruijn, and J. W. Newton (eds.), *Nitrogen fixation: Hundred years after.* Gustav Fischer, Stuttgart, West Germany.

Okon, Y., Y. Kapulnik, and S. Sarig. 1988b. Field inoculation studies with *Azospirillum* in Israel. In: N. S. Subba Rao (ed.), *Biological nitrogen fixation: Recent developments.* Gordon and Breach, NY.

Pal, U. R., and H. S. Malik. 1981. Contribution of *Azospirillum* brasilense to the nitrogen needs of sorghum (*Sorghum bicolor* (L.) Moench.) in humid subtropics. *Plant Soil* 63:501–504.

Patriquin, D. G. 1978. Factors affecting nitrogenase activity (acetylene reducing activity) associated with excised roots of the emergent halophyte *Spartina alterniflora* Loisel. *Aquat. Bot.* 4:193–210.

Patriquin, D. G., J. Dobereiner, and D. K. Jain. 1983. Sites and processes of association between diazotrophs and grasses. *Canad. J. Microbiol.* 29:900–915.

Rai, S. N., and A. C. Gaur. 1982. Nitrogen fixation by *Azospirillum* spp. and effect of *Azospirillum lipoferum* on the yield and N uptake of wheat crop. *Plant Soil* 69:233–238.

Rai, S. N., and A. C. Gaur. 1988. Characterization of *Azotobacter* spp. and effect of *Azospirillum* and *Azotobacter* as inoculant on the yield and N-uptake of wheat crop. *Plant Soil* 109:131–134.

Rao, V. R., D. N. Nayak, P.B.B.N. Charyulu, and T. K. Adhya. 1983. Yield response of rice to root inoculation with *Azospirillum*. *J. Agric. Sci.* 100:689–691.

Rao, V., P. K. Jena, and T. K. Adhya. 1987. Inoculation of rice with nitrogen fixing bacteria—problems and perspectives. *Biol. Fertil. Soils* 4:21–26.

Raverkar, K. P., and B. K. Konde. 1988. Effect of *Rhizobium* and *Azospirillum lipoferum* inoculation on the nodulation, yield and nitrogen uptake of peanut cultivars. *Plant Soil* 106:249–252.

Reinhold, B., T. Hurek, E.-G. Niemann, and I. Fendrik. 1986. Close association of *Azospirillum* and diazotrophic rods with different root zones of kallar grass. *Appl. Envir. Microbiol.* 52:520–526.

Rennie, R. J. 1980. ^{15}N-isotope dilution as a measure of dinitrogen fixation by *Azospirillum brasilense* associated with maize. *Canad. J. Microbiol.* 58:21–24.

Rennie, R. J., and R. I. Larson. 1979. Dinitrogen fixation associated with disomic chromosome substitution lines of spring wheat. *Canad. J. Bot.* 57:2771–2775.

Reynders, L., and K. Vlassak. 1982a. Use of *Azospirillum brasilense* as biofertilizer in intensive wheat cropping. *Plant Soil* 66:217–223.

Reynders, L., and K. Vlassak. 1982b. Physio-ecological aspects and agricultural importance of *Azospirillum*-plant root associations. In: W. Klingmuller (ed.), *Azospirillum: genetics, physiology, ecology*. Birkhauser, Basel.

Sadasivan, L., and C. A. Neyra. 1985. Cysts of Azospirilla under various cultural conditions. In: W. Klingmuller (ed.), *Azospirillum III: genetics, physiology, ecology*. Springer-Verlag, Berlin.

Sarig, S., A. Blum, and Y. Okon. 1988. Improvement of the water status and yield of field-grown grain sorghum *(Sorghum bicolor)* by inoculation with *Azospirillum brasilense*. *J. Agric. Sci. Camb.* 110:271–277.

Sarig, S., Y. Kapulnik, J. Nur, and Y. Okon. 1984. Response of non-irrigated *Sorghum bicolor* to *Azospirillum* inoculation. *Expl. Agric.* 20:59–66.

Sarig, S., Y. Kapulnik, and Y. Okon. 1986. Effect of *Azospirillum* inoculation on nitrogen fixation and growth of several winter legumes. *Plant Soil* 90:335–342.

Schank, S. C., K. L. Weier, and I. C. MacRae. 1981. Plant yield and nitrogen content of a digitgrass in response to *Azospirillum* inoculation. *Appl. Envir. Microbiol.* 41:342–345.

Schmidt, W., P. Martin, S. H. Omay, and F. Bangerth. 1988. Influence of *Azospirillum brasilense* on nodulation of legumes. In: W. Klingmuller (ed.), *Azospirillum IV: genetics, physiology, ecology*. Springer-Verlag, Berlin.

Shimshick, E. J., and R. R. Herbert. 1979. Binding characteristics of N_2-fixing bacteria to cereal roots. *Appl. Envir. Microbiol.* 38:447–453.

Smith, R. L., J. H. Bouton, S. C. Schank, K. H. Quesenberry, M. E. Tyler, J. R. Milam, M. H. Gaskins, and R. C. Littel. 1976. Nitrogen fixation in grasses inoculated with *Spirillum lipoferum*. *Science* 193:1003–1005.

Smith, R. L., J. H. Bouton, S. C. Schank, and K. H. Queensberry. 1977. Yield increases of tropical grain and forage grasses after inoculation with *Spirillum lipoferum* in Florida. In: A. Ayanaba and P. J. Dart (eds.), *Biological nitrogen fixation in farming systems of the tropics*. Wiley, NY.

Smith, R. L., S. C. Schank, and R. C. Littrell. 1984a. The influence of shading on associative N_2 fixation. *Plant Soil* 80:43–52.

Smith, R. L., S. C. Schank, J. R. Milam, and A. A. Baltensperger. 1984b. Responses of *Sorghum* and *Pennisetum* species to the N_2-fixing bacterium *Azospirillum brasilense*. *Appl. Envir. Microbiol.* 37:1331–1336.

Subba Rao, N. S., K. V. Tilak, C. S. Singh, and M. Lakshmi-Kumari. 1979. Response of a few economic species of graminaceous plants to inoculation with *Azospirillum lipoferum*. *Curr. Sci.* 48:133–134.

Sumner, M. E. 1978. Interpretation of nutrient ratios in plant tissue. *Comm. Soil Sci. Plant Anal.* 9:335–345.

Taylor, R. W. 1979. Response of two grasses to inoculation with *Azospirillum* spp. in a Bahamian soil. *Trop. Agric.* 56:361–365.

Tien, T. M., M. H. Gaskens, and D. H. Hubbell. 1979. Plant growth substances produced by *Azospirillum brasilense* and their effect on the growth of pearl millet (*Pennisetum americanum* L.). *Appl. Envir. Microbiol.* 37:1016–1024.

Tilak, K. V., and N. S. Subba Rao. 1987. Association of *Azospirillum brasilense* with pearl millet (*Pennisetum americanum*). *Biol. Fertil. Soils* 4:97–102.

Umali-Garcia, M., D. H. Hubbell, M. H. Gaskens, and F. B. Dazzo. 1980. Association of *Azospirillum* with grass roots. *Appl. Envir. Microbiol.* 39:219–226.

Umali-Garcia, M., D. H. Hubbell, M. H. Gaskens, and F. B. Dazzo. 1981. Adsorption and mode of entry of *Azospirillum brasilense* to grass roots. In: P. B. Vose and A. P. Ruschel (eds.), *Associative N_2 fixation*. CRC Press, Boca Raton, FL.

van Berkum, P., and B. B. Bohlool. 1980. Evaluation of nitrogen fixation by bacteria in association with roots of tropical grasses. *Microbiol. Rev.* 44:491–517.

Vance, C. P. 1983. *Rhizobium* infection and nodulation: a beneficial plant disease. *Ann. Rev. Microbiol.* 37:399–424.

Venkateswarlu, B., and A. V. Rao. 1983. Response of pearl millet to inoculation with different strains of *Azospirillum brasilense*. *Plant Soil* 74:379–386.

Watanabe, I., and C. Lin. 1984. Response of wetland rice to inoculation with *Azospirillum lipoferum* and *Pseudomonas* sp. *Soil Sci. Plant Nutr.* 30:117–124.

Wani, S. P., S. Chandrapalaiah, and P. J. Dart. 1985. Response of pearl millet cultivars to inoculation with nitrogen fixing bacteria. *Expl. Agric.* 21:175–182.

Warembourg, F. R., R. Dressen, K. Vlassak, and F. Lafont. 1987. Peculiar effect of *Azospirillum* inoculation on growth and nitrogen balance of winter wheat (*Triticum aestivum*). *Biol. Fertil. Soils* 4:55–59.

Whallon, J. H., G. Acker, and H. El-Khawas. 1985. Electron microscopy of young wheat roots inoculated with *Azospirillum*. In: W. Klingmuller (ed.), *Azospirillum III: genetics, physiology, ecology*. Springer-Verlag,. Berlin.

Yahalom, E., Y. Kapulnik, and Y. Okon. 1984. Response of *Setaria italica* to inoculation with *Azospirillum brasilense* as compared to *Azotobacter chroococcum*. *Plant Soil* 82:77–85.

Yahalom, E., Y. Okon, and A. Dovrat. 1987. *Azospirillum* effects on susceptibility to *Rhizobium* nodulation and on nitrogen fixation of several forage legumes. *Canad. J. Microbiol.* 33:510–514.

Yahalom, E., Y. Okon, and A. Dovrat. 1988. Early nodulation in legumes inoculated with *Azospirillum* and *Rhizobium*. *Symbiosis* 6:69–80.

Zambre, M. A., B. K. Konde, and K. R. Sonar. 1984. Effect of *Azotobacter chroococcum* and *Azospirillum brasilense* inoculation under graded levels of nitrogen on growth and yield of wheat. *Plant Soil* 79:61–67.

Application of ^{57}Fe Mössbauer Spectroscopy to Problems in Clay Mineralogy and Soil Science: Possibilities and Limitations

E. Murad*

I. Introduction

The recoil-free emission and resonant absorption of γ rays (Mössbauer spectroscopy) has, since its discovery three decades ago (Mössbauer, 1958a, 1958b), become an increasingly important technique for the study of iron-bearing minerals. The fundamental requirement for the observation of γ-ray resonance is an inhibition of recoil in both emitting and absorbing atoms, so the development of a Mössbauer spectrum requires these to be bound in solids. Because resonant interactions between γ rays and matter take place at nuclei, Mössbauer spectroscopy is sensitive to the immediate nuclear environment in the γ-ray source and absorber. This is therefore a typical short-range method, which may serve as a probe for the electric and magnetic conditions in the vicinity of nuclei. A

*Lehrstuhl für Bodenkunde, Technische Universität München, D-8050 Freising-Weihenstephan, Federal Republic of Germany.

consequence of this is that Mössbauer spectroscopy can be successfully applied to the study of materials of particle sizes so small that the application of other, "classic" techniques such as electron microscopy or X-ray diffraction is considerably impeded.

Of about 100 isotopes that show a Mössbauer effect, only ^{57}Fe has been used to any significant extent in clay mineralogy and soil science. This chapter is therefore concerned exclusively with ^{57}Fe. ^{57}Fe has an acceptably high natural abundance (2.14%), the source emitting the 14.4-keV radiation used for ^{57}Fe Mössbauer spectroscopy (^{57}Co) has a reasonably long half-life (270 d), and the 14.4-keV resonance has a narrow linewidth (0.097 mm/s). Since Mössbauer spectroscopy is an element-specific technique, elements other than that under survey cause only a dilution of the isotope under study, but no matrix effects. All of these factors add up to make ^{57}Fe Mössbauer spectroscopy a very useful technique for the study of iron-bearing minerals.

Iron oxides (see Kistner and Sunyar, 1960; Hrynkiewicz and Kulgawczuk, 1963) and clay minerals (Weaver et al., 1967) were studied relatively early by Mössbauer spectroscopy. Most of this earlier work on minerals, however, was concerned with the accumulation of type data. The first systematic studies of minerals formed in the ambient environment making use of Mössbauer spectroscopy were carried out by Janot et al. (1968, 1971, 1973); studies concerned directly with soils were first published by Yassoglou et al. (1972) and Gangas et al. (1973). The potency of Mössbauer spectroscopy having been realized, numerous papers concerned with the application of this method to soil and clay mineral research have since been published. Improved instrumentation and evaluation techniques have facilitated the application of Mössbauer spectroscopy and led to the publication of increasingly more detailed and precise data on iron oxides and clay minerals during recent years.

Various textbooks exist in which the physical background of Mössbauer spectroscopy is described in considerable detail. The basic principles of Mössbauer spectroscopy are therefore outlined below in just sufficient detail to allow readers who are not familiar with this technique to follow the argumentation. The existence of the periodical *Mössbauer Effect Reference and Data Journal,* in which practically all publications making use of the Mössbauer effect are listed, has proven particularly useful. Several reviews on the application of Mössbauer spectroscopy to related fields have also been recently published. The Mössbauer spectra of silicates in general have been described by Coey (1984), and clay minerals in particular by Coey (1980), Goodman (1980), and Heller-Kallai and Rozenson (1981). The application of Mössbauer spectroscopy to the study of iron oxides has recently been reviewed by Bowen and Weed (1981, 1984), Murad and Johnston (1987), and Murad (1988).

The purpose of this contribution is to outline the possibilities Mössbauer spectroscopy offers for research in soil science. In particular, it is intended to demonstrate where the application of Mössbauer spectroscopy can be recommended and where not, to suggest fields where this technique holds promise in future work, and to point out possible complications and pitfalls. Minerals that are of relevance to soil studies have been subdivided into two main groups: those that are paramagnetic at least down to 77 K (silicates, carbonates, green rusts) or diamagnetic (sulfides), and those that *may* show magnetic order at or above 77 K (iron oxides). For reasons of practicality, the latter section also includes minerals that may be superparamagnetic at temperatures significantly below 77 K (such as ferrihydrite).

A moderate amount of data is available on the Mössbauer spectra of organic matter from soils and sediments. Extraction procedures are often necessary to distinguish the contributions of organically bound iron from those of inorganic minerals, and it has been indicated that these procedures may change the forms in which iron is bound (Dickson et al., 1979). The extracts themselves may be so iron-poor that subsequent doping is necessary to enhance Mössbauer absorption. On the whole, in-depth information on the Mössbauer spectra of organic matter *as it occurs* in soils is still lacking, and this subject will therefore not be touched on further here.

II. Physical Background and Instrumentation

If the nuclear environments in a Mössbauer source and absorber differ (which is normally the case), resonance will require a shift of the γ-ray energy. This energy shift is generally accomplished by moving the source, making use of the Doppler effect. To record a spectrum in the (commonly used) transmission geometry, the source is moved over a range of velocities, and the γ radiation transmitted through the absorber is recorded as a function of velocity (i.e., energy). A Mössbauer spectrum thus consists essentially of a set of radiation counts versus velocity data, and the energies at which resonances occur are given as source velocities in millimeters per second.

The nuclear energy levels of ^{57}Fe and transitions between these which lead to the development of Mössbauer spectra are shown in Figure 1. The nuclear energy levels may vary slightly or be split into sublevels as a result of the so-called hyperfine interactions between the nuclei and electrons surrounding them. Quantum theory determines which transitions between the individual energy levels are possible. The decay scheme of ^{57}Co to ^{57}Fe and the γ rays emitted by this during the transition to the ground state are shown in the inset of Figure 1. Only the γ rays emitted during the transition from the $I = \frac{3}{2}$ excited state to the ground state ($I = \frac{1}{2}$) have an energy that is low enough (14.4 keV) to be of importance for routine Mössbauer spectroscopy.

Parameters which can be derived from Mössbauer spectra and used to characterize the absorber are the isomer shift δ, quadrupole splitting ΔE_Q, magnetic hyperfine field B_{hf}, and linewidths at half maximum W. The isomer shift is given by the center of gravity, and the quadrupole splitting by the separation of the two lines of a doublet (Figure 1). The magnetic hyperfine field of magnetically ordered materials can be derived from the splitting of a sextet, and the isomer shift and quadrupole splitting of such a substance from the center of gravity and difference in splitting of the outer line pairs (Figure 1). Doppler shifts for the observation of resonance require velocities up to about ± 12 mm/s.

The isomer shift is directly proportional to the s-electron density at the

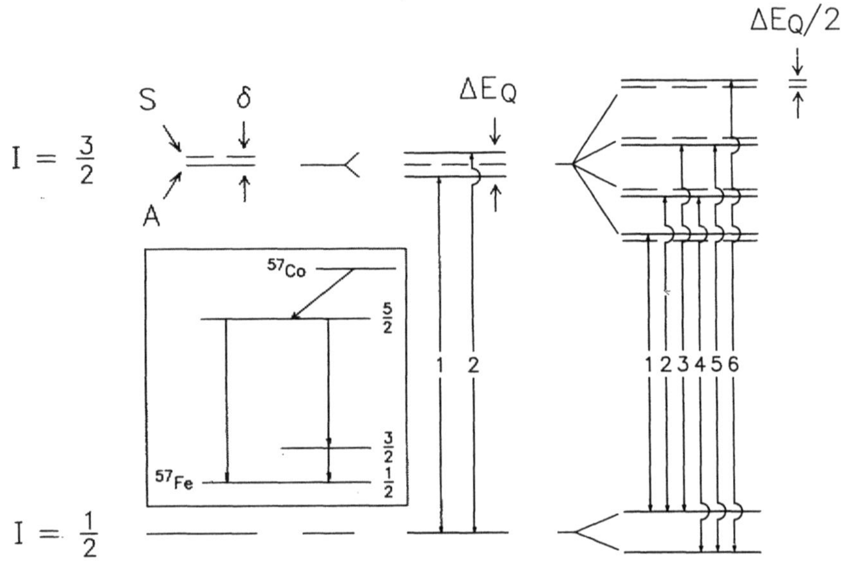

Figure 1. Nuclear energy levels of ^{57}Fe in the excited (I = 3/2) and ground (I = 1/2) states. Transitions between these are shown for the cases of a noncubic electric field, giving a quadrupole-split Mössbauer doublet, and in the presence of a nonvanishing magnetic field, giving a sextet. Energy shifts betwen source and absorber δ, between the two sublevels in the I = 3/2 state leading to the development of a quadrupole splitting ΔE_Q, and the perturbation due to a quadrupole intraction superimposed upon the magnetic hyperfine splitting $\Delta E_Q/2$ are also indicated. Note that the diagram is not to scale; the energy difference between the I = 3/2 excited and the ground states may be over 10^{11} times higher than the splitting of the substates. The inset shows the decay scheme for ^{57}Co and the γ rays emitted during the transition of ^{57}Fe thus formed to the ground state.

nucleus, and can be used as an indicator for the oxidation state and coordination of iron. The values of isomer shifts actually observed depend on the source matrix material (usually a cubic metal into which ^{57}Co has been diffused). Tables for the recalculation of isomer shifts measured with different sources have been published (see Stevens, 1983); these are usually reported relative to metallic iron (δ/Fe). Ranges of isomer shifts for divalent and trivalent iron in different coordinations have been published by Bancroft et al. (1967). For high-spin trivalent iron in octahedral coordination this is typically about 0.35 mm/s relative to metallic iron, whereas high-spin divalent iron has a significantly higher isomer shift centered around approximately 1 mm/s.

In paramagnetic minerals (e.g., all iron-bearing silicates at room temperature and nearly all at 77 K) with a noncubic symmetry at the nucleus, the excited state (nuclear spin $I = \frac{3}{2}$) is split into two sublevels by an electric field gradient. Two transitions between the excited state and the degenerate ground state ($I = \frac{1}{2}$) are therefore possible. Ideally this results in a distinct two-line absorption (quadrupole-split doublet) for each iron site and valence.

Quadrupole splittings can be divided into a valence contribution arising from the electrons of the iron atom involved in resonant absorption, and a lattice contribution arising from the external ligand charge distribution. High-spin Fe(III) has five unpaired 3*d* electrons, each located in a different orbital. This results in a cubically symmetric charge distribution and reduces the valence contribution to the quadrupole splitting to zero. For compounds of high-spin Fe(III), quadrupole splittings therefore depend essentially on the site symmetry, and may vary from zero (for cubic symmetry) to >1 mm/s for highly distorted coordinations. For high-spin Fe(II) in a coordination that is distorted from cubic, the quadrupole splitting has both valence and lattice contributions, and may vary strongly as a function of temperature.

In many iron oxides, magnetic hyperfine fields act at the nuclei even at room temperature, so that magnetic hyperfine splitting is of particular importance in iron oxide mineralogy. In iron oxides that are paramagnetic at room temperature and some iron-rich silicates, magnetic hyperfine splitting can be induced by cooling down to 77 K using liquid nitrogen or, if necessary, to 4.2 K using liquid helium. Under a nonvanishing magnetic field the energy levels in the $I = \frac{3}{2}$ state are raised from degeneracy and split into four different sublevels (Zeeman splitting), and those in the ground state split into two sublevels. There are six "allowed" transitions between the $I = \frac{3}{2}$ and $I = \frac{1}{2}$ states, so that the Mössbauer spectrum of a magnetically ordered material consists of a sextet. In the case of a noncubic crystal field, a quadrupole interaction may be superimposed upon this magnetic hyperfine splitting. For high-spin Fe(III) the quadrupole interaction is much weaker than the magnetic splitting, and may therefore be considered a perturbation of this (see Figure 1). For high-spin Fe(II), however, both the magnetic dipole and the electric quadrupole interactions may be of the same order of magnitude, which can complicate the appearance and evaluation of such spectra considerably.

The mathematical function describing resonance, and therefore the Mössbauer line shape, is the Lorentzian. Linewidths for the ^{57}Fe resonance are theoretically 0.097 mm/s each for the source and the absorber; that is, the minimum total linewidth for this resonance is 0.195 mm/s. In practice, linewidths of <0.3 mm/s are usually considered acceptable. Significantly wider lines or noticeable deviations from the Lorentzian shape indicate either inadequate experimental conditions, relaxation phenomena, or more than one iron site with similar, but not completely coincident parameters.

Line intensities depend on various factors, among which are the spectrometer geometry, the recoil-free fraction (which is an inherent sample property), and the iron content of the Mössbauer absorber. Because linewidths increase with the iron content of samples, a compromise is usually necessary between an acceptable line broadening due to finite sample thickness and a strong signal. Iron contents of about 5 mg/cm^2 will usually give satisfactory results.

Numerous commercial Mössbauer spectrometers are available today. All spectrometers comprise a driving system which moves the source over the required velocity range, a radiation detector, a single-channel analyzer to sort out the γ radiation used for Mössbauer spectroscopy from the rather complex spectrum emitted by the source, and a multichannel analyzer or microprocessor in which the radiation versus velocity data can be stored. A laser interferometer is sometimes used for velocity calibration. Cryostats (or, less frequently in the line of research we are concerned with, ovens) will be necessary if measurements are to be made at temperatures other than room temperature. Occasionally Mössbauer spectra may be recorded in the scattering mode, in which the intensity of conversion electrons or the radiation emitted during decay of the excited state following resonance in the absorber is registered as a function of source velocity. Because scattered conversion electrons and X rays emitted during decay of the $I = \frac{3}{2}$ excited state of ^{57}Fe have maximum penetration depths of about 0.1 and 20 μm, respectively, this technique is particularly suited to the study of surfaces or thin coatings on other materials.

Computer analysis of Mössbauer data is indispensable. Some modern units use personal computers to do this job in addition to controlling the Mössbauer unit. Computer analysis may involve some or all of the following steps: folding of spectra that consist of mirror halves, fitting of Lorentzian singlets, doublets, or sextets to the spectra, and plotting data points, fit envelopes, individual peaks, and residuals.

Because the intensities of Mössbauer lines are orientation dependent, care must be taken in clay mineral studies to reduce orientation to a minimum. Techniques for absorber preparation and computer programs used for the evaluation of Mössbauer data vary widely. A recent survey involving seven different laboratories in which minerals are routinely studied by Mössbauer spectroscopy nevertheless showed interlaboratory differences to be surprisingly small (Dyar, 1984).

III. Paramagnetic and Diamagnetic Minerals

A. Clay Minerals

No clay mineral is magnetically ordered at room temperature or 77 K. Few, very iron-rich clay minerals may order magnetically below 20 K, but most iron-containing clay minerals remain paramagnetic down to 4.2 K. Information which has been derived from Mössbauer spectra of paramagnetic clay minerals is the site coordination, the degree of occupation of different sites, and the oxidation state of iron. Because different oxidation states of iron lead to distinctly different Mössbauer isomer shifts, a distinction of high-spin Fe(II) and Fe(III) is usually relatively straightforward. Although it is possible to calculate average Mössbauer parameters for different clay minerals (Pollak and Stevens, 1986), it must be noted that, for clays, Mössbauer parameters may vary considerably between different samples of the same mineral (Heller-Kallai and Rozenson, 1981). Mössbauer spectroscopy therefore has only a limited value as a technique for the *identification* of clay minerals, and certainly should not be used as a substitute for X-ray diffraction.

In the simplest conceivable case, the Mössbauer spectrum of a paramagnetic clay mineral containing Fe(III) in only one site of the octahedral sheet should consist of a doublet with an isomer shift in the range commonly encountered for octahedrally coordinated Fe(III) and a quadrupole splitting which depends on the distortion of the site from cubic symmetry. For Fe(II) analogous Mössbauer spectra with a higher isomer shift than for Fe(III) will be developed; a prediction of the quadrupole splitting for Fe(II) is rather complicated because of the valence contribution to the electric field gradient (Goodman, 1976). In practice, deviations from such idealized models can be expected because of local fluctuations of composition or structural defects. In minerals with multiple cation sites, a separate doublet should be developed for each site occupied by iron. The resolution of such doublets may, however, be difficult or even impossible if the sites are similar. It must furthermore be ensured that only the clay mineral under survey contributes to the Mössbauer spectrum, that is, that a clay mineral sample contains no other iron-bearing constituents. This is a premise that is often made but rarely verified (see below).

The environment of an atom is determined not only by its position in a crystal lattice, but to some extent also by its neighbors, next-nearest neighbors, and so on. Isomorphous substitutions are common in the octahedral sheets of both dioctahedral and trioctahedral clay minerals. As a consequence of such substitutions, the environments of atoms located in the octahedral sheet will no longer be well defined but rather may, depending on the extent of variation of neighboring atoms, be "smeared out." Depending on whether variations of the environment of iron atoms in the octahedral sheet occur at random or in an ordered manner, these variations will lead to either line broadening or the development of more than one discrete component in Mössbauer spectra. Thus iron was observed to occupy only one site in dioctahedral 1:1 phyllosilicates (although broad lines suggest some site variability), but the occupation of more than one site was indicated in trioctahedral samples (Rozenson et al., 1979). Fysh et al. (1983), in contrast, suggested that significant portions of the Fe(III) content of kaolins could, rather than constituting part of the kaolinite structure, be due to associated iron oxides of very small particle size, whereas structural Fe(III) in kaolinites should usually be low enough to result in paramagnetic relaxation (which could account for at least part of the line broadening observed at room temperature; see Figure 2).

Other dioctahedral clay minerals may have much more complicated Mössbauer spectra. Of these, smectites and in particular nontronite have received considerable attention lately. Mössbauer spectra of nontronite are commonly fitted with three Fe(III) doublets (Goodman et al., 1976), although four doublets (Johnston and Cardile, 1985) or even distributions of quadrupole-split doublets (Murad, 1987) could be more appropriate (Figure 3). To complicate things even

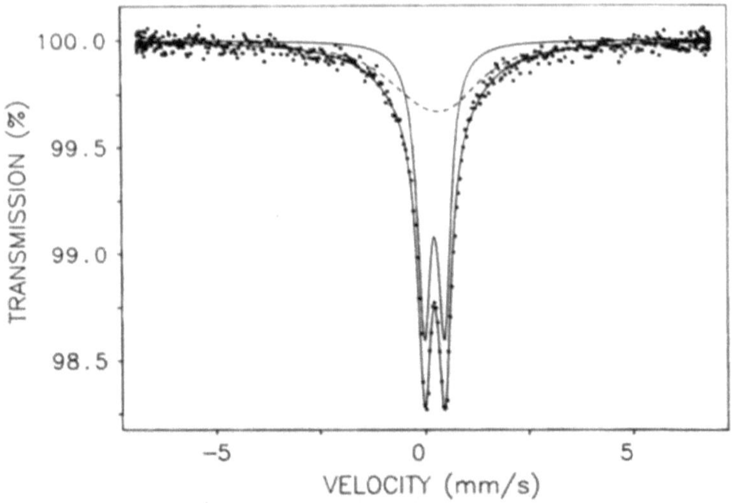

Figure 2. Mössbauer spectra of a kaolinite at room temperature fitted with an Fe(III) doublet plus a broad magnetic resonance to account for paramagnetic relaxation (broken subspectrum). (Data courtesy of U. Wagner.)

more, nontronite appears to be frequently associated with iron oxides that are superparamagnetic at room temperature (see next section), and these give resonant peaks which are unresolvably superimposed upon the nontronite spectra (Murad, 1987). The determination of physically meaningful parameters for such mineralogically complex samples requires Mössbauer spectra to be taken at different temperatures and over different velocity ranges, and thus may be much more laborious than is apparent at first sight.

Tetrahedrally coordinated iron can be distinguished from octahedrally coordinated iron by a lower isomer shift, and Fe(II) can be distinguished from Fe(III) by a higher isomer shift and quadrupole splitting. This can be used to estimate the Fe(II) and Fe(III) populations on different sites in clay minerals. The validity of such determinations is based on the premise that the recoil-free fractions are independent of site coordination and oxidation state. Because the recoil-free fractions vary as a function of lattice vibrations, it may be necessary to carry out such measurements at very low temperatures in order to minimize vibration-dependent variations. The application of an appropriate computer fit model may require a detailed knowledge of the sample composition (Goodman, 1976). Finally, for thus determined Fe(II)/Fe(III) ratios and site populations to have any significance, interferences by associated minerals must—as for the determination of accurate Mössbauer parameters—be excluded.

What, then, are possible uses of Mössbauer spectroscopy in clay mineralogy?

An inherent problem in clay mineralogy is that, because of the small particle sizes, it is difficult to obtain pure, monomineralic samples. Provided such samples are available for study, Mössbauer spectroscopy can be very useful to

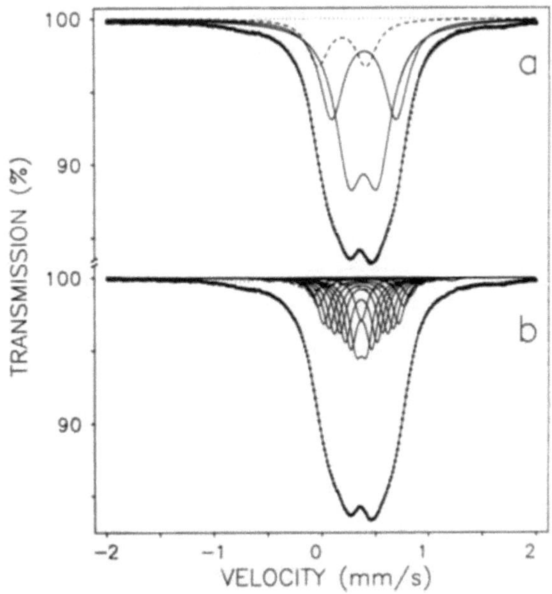

Figure 3. Mössbauer spectra of a nontronite at 77 K. (a) Fitted with four doublets: two doublets (solid subspectra) for the octahedral iron resonance plus one doublet each for tetrahedral iron (broken subspectrum) and the inner two lines of a goethite impurity (dotted line). (b) Fitted with a distribution of quadrupole-split doublets for the octahedral iron resonance and (again) one doublet each for tetrahedral iron and the inner two goethite lines.

determine which lattice sites are occupied by iron, the distortions of these sites, and the oxidation state of iron. Because iron oxides usually order magnetically at higher temperatures than clay minerals, Mössbauer spectra taken at low temperatures can help distinguish between oxidic and silicate iron. This may allow crystallographic work on clay samples of complex mineralogy to be done without prior removal of iron oxides by chemical treatments (Murad, 1987).

Early work by Bowen et al. (1969) showed the Fe(III) proportion in biotites to increase following K depletion, but the oxidation of iron left the isomer shifts and quadrupole splittings essentially unaffected. Using Mössbauer spectroscopy, Yassoglou et al. (1972) and Goodman and Wilson (1973) observed natural weathering of biotites to be accompanied by a noticeable oxidation of iron too. Rozenson and Heller-Kallai (1976a, 1976b) studied the reduction of Fe(III) in dioctahedral smectites using hydrazine, Na dithionite (Figure 4), and Na_2S. The extent and reversibility of reduction were observed to vary as a function of both the reducing agent and the mineralogy of the sample. Oxidation of Fe(II) in montmorillonites, on the other hand, was quite readily effected (Rozenson and Heller-Kallai, 1978), indicating that chemical methods of determining the CEC can also affect the degree of oxidation of structural iron.

Thus the oxidation, reduction, and decomposition of clay minerals both in the

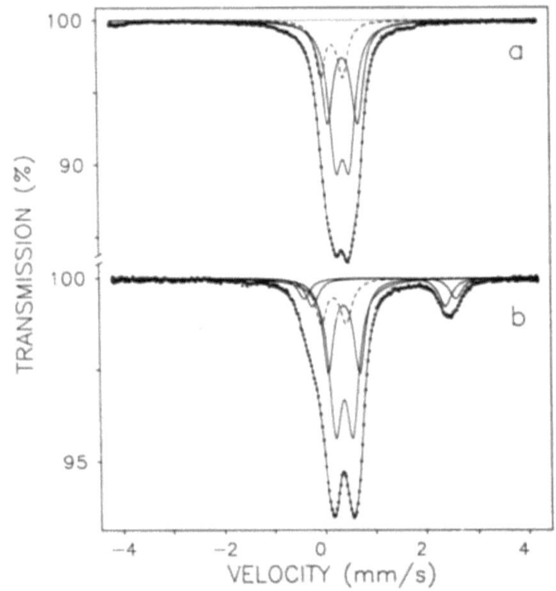

Figure 4. Mössbauer spectra of a nontronite at 77 K. (a) Before and (b) after reduction with Na dithionite. Subspectra as for Figure 3.

laboratory and in natural weathering sequences can be monitored relatively reliably using Mössbauer spectroscopy.

Mössbauer spectra of Fe(II) exchanged smectites show an abrupt onset of resonant Fe(II) absorption at about 210 K, with a quadrupole splitting that is higher than that of Fe(II) in structural sites (Diamant et al., 1982; Helsen and Goodman, 1983). This indicates the presence of solvated Fe(II) in the interlayer, freezing of which results in an abrupt onset of recoil-free absorption, and allows a distinction to be made between interlayer and structural Fe(II).

Irreversible oxidation of Fe(II) contained in clay minerals may lead to the formation of iron oxides, which can be identified by Mössbauer spectroscopy even at relatively low concentrations (although low-temperature spectra may be necessary to distinguish these from structural iron).

B. Carbonates and Sulfides

In siderite, $FeCO_3$, iron is in the high-spin divalent state, that is, it has four unpaired and two paired 3*d* electrons. Pure siderite is paramagnetic down to its Néel temperature of 38 K, below which it is magnetically ordered. Mössbauer spectra of paramagnetic siderite may have an asymmetric appearance due to preferred particle orientation [Figure 5(a)]. The quadrupole splitting increases inversely with temperature, rising from 1.78 mm/s at room temperature to over 2 mm/s near 40 K. Neither the quadrupole splitting nor the isomer shift (δ/Fe $\approx$ 1.2

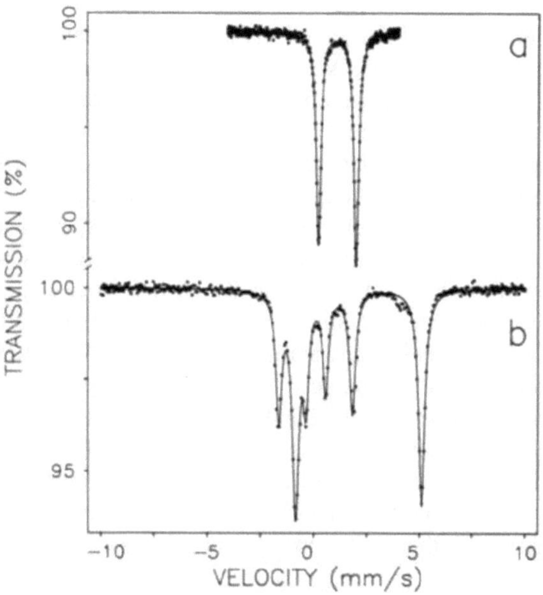

Figure 5. Mössbauer spectra of a siderite. (a) At room temperature, showing asymmetric peak development. (b) At 4.2 K, showing magnetic order. In spectrum (b) the relatively high quadrupole interaction has shifted peak 1 of Figure 1 to a higher velocity than peak 2.

mm/s at room temperature) seem to vary distinctively with siderite composition (Amigó and Fortuné, 1981), so that these cannot be used to characterize this mineral. Fe(II) occurring in ferriferous magnesites, dolomites, or calcites has an essentially similar Mössbauer behavior (DeGrave and Vochten, 1985), although the intensity of the resonance will, of course, be reduced with a decreasing iron content.

Because weathering of siderite and related carbonates is accompanied by an oxidation of Fe(II), this will be immediately apparent in Mössbauer spectra. Mössbauer spectroscopy has been used to characterize siderite decomposition products (Eymery et al., 1978) and could be of use to delineate ferroan carbonate mineral weathering.

The two most common iron sulfides, pyrite and marcasite, differ from the iron-bearing minerals described above by containing Fe(II) in the low-spin state. Low-spin Fe(II) has six paired 3*d* electrons and is therefore diamagnetic. These minerals therefore do not order magnetically at any temperature. Because of the covalent bonding character, the isomer shifts are significantly lower than for high-spin Fe(II), and resemble those of high-spin Fe(III). Although Mössbauer parameters of natural pyrites and marcasites (δ/Fe = 0.313(8) and 0.277(2) mm/s, ΔE_Q = 0.611(3) and 0.503(7) mm/s, respectively, at room temperature) vary somewhat, probably due to chemical impurities, these minerals can be unambiguously distinguished by Mössbauer spectroscopy (Evans et al., 1982).

The Mössbauer spectra of pyrite and marcasite thus resemble those of many Fe(III)-containing clay minerals, which may complicate a distinction between these. Vértes et al. (1981) suggested firing pyrite–biotite mixtures at 900°C in air to distinguish the contributions of these iron-bearing minerals to the Mössbauer spectra: while pyrite is converted to (magnetically ordered) hematite, biotite remains paramagnetic at room temperature. This technique is, however, not expected to work for sulfide–clay mineral associations, because most clay minerals will break down on firing to such high temperatures and also form hematite (Simopoulos et al., 1975; Heller-Kallai and Rozenson, 1980).

Because pyrite and marcasite are common constituents of sulfidic ores and coal, the weathering of these minerals in the ambient environment is of prime importance. Iron sulfates and oxides, which are the typical products of iron sulfide weathering, can be identified and characterized by Mössbauer spectroscopy even at low overall concentrations. Mössbauer spectroscopy has been found suitable to monitor environmental sulfide oxidation (Huggins et al., 1980) and therefore bears considerable potential as a technique to characterize the oxidation status and potential weathering hazard of mine dumps.

C. Green Rusts

The green rusts are a family of metastable Fe(II)Me(III) hydroxy salts, laboratory syntheses of which are relatively straightforward (Taylor, 1980; Taylor and McKenzie, 1980). The presence of Fe(II) renders these compounds very susceptible to oxidation, so that special precautions are required for their preservation. The observation of transient greenish phases in hydromorphic soils has been taken as an indication that green rusts might be constituents of these. The instability of green rusts, however, is adverse to their identification by standard mineralogical techniques such as X-ray diffraction, so that a positive identification of green rusts in soils has not been effected so far.

Mössbauer spectra of paramagnetic Fe(II)Al and Fe(II)Fe(III) hydroxycarbonate green rusts synthesized in the laboratory have been shown to comprise two Fe(II) doublets and one or two Fe(III) doublets, respectively (Murad and Taylor, 1984). The outer of the two Fe(II) doublets has narrow lines (W = 0.28 mm/s), indicating Fe(II) to be bound in an ordered environment. Doublets with essentially the same parameters (δ/Fe = 1.15, ΔE_Q = 2.88 mm/s) were also observed in samples from other syntheses (Murad and Taylor, 1986). Although these parameters do not allow a positive distinction from some trioctahedral clay minerals, the relatively rapid disappearance of this component upon exposure of the sample to the atmosphere (Figure 6) could be used as a criterion for green rust identification in mineralogically complex samples (Murad and Taylor, 1986). The fact that taking Mössbauer spectra at low temperatures (such as by cooling the absorber with liquid nitrogen) would essentially inhibit unintentional oxidation of green rusts, makes this the most promising technique for the identification of such compounds in soil environments, should they exist there.

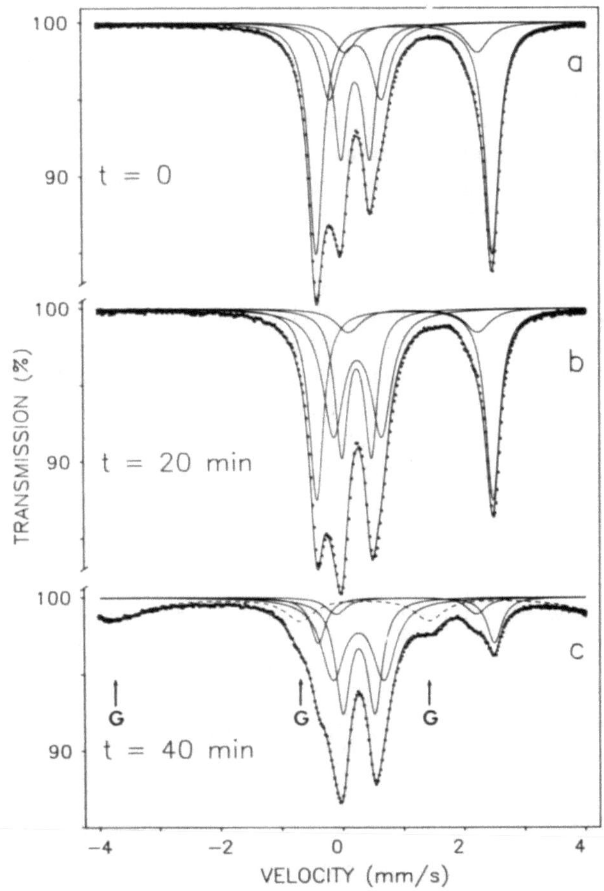

Figure 6. Mössbauer spectra of an Fe(II)Fe(III) hydroxycarbonate green rust at 120 K. (a) As prepared. (b) After 20-min exposure to atmosphere. (c) After 40-min exposure to atmosphere. Spectra (a) and (b) are fitted with two Fe(II) and two Fe(III) doublets. Spectrum (c) has three additional lines fitting peaks (marked G) of magnetically ordered goethite (broken subspectrum). (Modified from Murad and Taylor, 1986).

IV. Magnetically Ordered Minerals

A. Iron Oxides*

Magnetic ordering temperatures for iron oxides range from 955 K for hematite to 77 K for lepidocrocite. In the magnetically ordered state, the iron oxides described here can be antiferromagnetically (antiparallel spins with equal mo-

*In the context of this review, the term *iron oxide* includes the various oxyhydroxides (FeOOH) and ferrihydrite.

ments that cancel out) or ferrimagnetically ordered (antiparallel spins with unequal moments that do not cancel out completely). The antiferromagnetic ordering temperature is usually called the Néel temperature T_N. Ferro- and ferrimagnetic ordering takes place at the Curie temperature T_C. Between T_C (955 K) and the so-called Morin temperature (260 K) hematite shows a distinctive magnetic behavior: although the spin arrangement is essentially antiferromagnetic, the spins are not exactly antiparallel, so that a weak net ferromagnetic interaction ensues.

Above their respective magnetic ordering temperatures all iron oxides are paramagnetic with Mössbauer spectra that consist of rather unspecific doublets. Below the ordering temperatures magnetic hyperfine fields acting at the nuclei lead to the development of magnetic splitting. Both the magnetic ordering temperatures and the hyperfine fields at different temperatures may be diagnostic for the diverse iron oxides. In Table 1 magnetic ordering temperatures and magnetic hyperfine fields at temperatures frequently used for Mössbauer

Table 1. Magnetic properties of iron oxides that have been reported to occur in soils

				Hyperfine field (T)		
Mineral	Formula	T_N, T_C (K)	Magnetic character[a]	295 K	77 K	4.2 K
Hematite	α-Fe_2O_3	955	wf[b]	51.8	53.5	53.5
			af		54.2	54.2
Maghemite	γ-Fe_2O_3	c	fi	50.0	52.8	52.8
					52.4	52.4
Magnetite	Fe_3O_4	850	fi[d]	{49.0 / 46.0		
Goethite	α-FeOOH	400	af	38.2	50.0	50.6
Akaganeite	β-FeOOH	295	af[e]	{ —	47.3	48.9
				—	46.3	47.8
				—	43.7	47.3
Lepidocrocite	γ-FeOOH	77	af	—	—	45.8
Feroxyhite	δ'-FeOOH	440	fi	42.0	52.0	52.5
Ferrihydrite	~$Fe_5HO_8 \cdot 4H_2O$	f	sp	—	≤45.1	46.5–50

*Magnetic ordering temperatures (T_N for antiferromagnetic and T_C for ferro- and ferrimagnetic order) and hyperfine fields given for chemically pure, well-crystallized samples. [Modified after Murad (1988).]

[a]fi: ferri-; af: antiferro-; wf: weakly ferro-; sp: speromagnetic.

[b]Above the Morin transition (weakly ferromagnetic phase).

[c]Published values vary between 743 and 1020 K (Morrish, 1980).

[d]Spectra consist of two discrete components above and several discrete components below the Verwey transition (at about 119 K).

[e]Spectra show at least three discrete magnetic components.

[f]Complete magnetic order usually requires temperatures below 77 K.

measurements are given for those iron oxides which could be of relevance to soil studies.

In contrast to the magnetic behavior of well-crystallized iron oxides, small particle sizes can depress the onset of magnetic order to temperatures below those listed in Table 1. This phenomenon of *superparamagnetism* is a common feature among pedogenic iron oxides, which frequently form rather small particles. Because of its inherently small particle size, ferrihydrite does not have a well-defined ordering temperature but, depending on the sample crystallinity, orders over a range of more or less low temperatures. Poor crystallinity may also lead to a freezing of spin states in random directions in magnetically ordered ferrihydrite (*speromagnetic* behavior; Coey and Readman, 1973).

1. Room-Temperature Spectra

Room-temperature Mössbauer spectra are cheap to record because they require neither expensive coolants nor sophisticated cryogenic equipment. A further value lies in the fact that room-temperature spectra allow a ready distinction to be made between iron oxides that *can* be magnetically ordered at 295 K (hematite and goethite as the most common minerals, also magnetite and maghemite; see Table 1) and those that *cannot* possibly contribute to the magnetically ordered section of a spectrum because they have lower ordering temperatures.

Magnetic hyperfine fields show inverse trends with temperature, increasing very rapidly in the vicinity of the ordering temperature and gradually approaching saturation values near 0 K. Well-crystallized, pure hematite orders magnetically at 955 K and therefore has a "reduced" temperature (T/T_C) of about 0.31 at 295 K; goethite has a much higher T/T_N of 0.74 at the same temperature. It is thus obvious that goethite is further removed from saturation conditions at room temperature than hematite. This is reflected in the distinctly different hyperfine fields [38.2 versus 51.8 tesla (T)], which facilitates a distinction of these two minerals at room temperature. It is also the reason why the magnetic hyperfine field of goethite increases much more between 295 and 77 K than that of hematite (see Figure 7 and Table 1).

The magnetic ordering temperatures and hyperfine fields given in Table 1 are those for pure, well-crystallized samples. Crystalline imperfections and chemical substitutions of other elements for iron tend to both lower the ordering temperatures and reduce the saturation hyperfine fields. While at low reduced temperatures only changes in the saturation field will be noticeable, both the saturation field reduction and the lower ordering temperature are effective at higher reduced temperatures. Deviations from ideal behavior will therefore affect Mössbauer spectra of goethite more than those of hematite at room temperature.

The substitution of aluminum for iron in iron oxides is a widespread and well-documented phenomenon (Schwertmann, 1985). Al(III) has the same electronic configuration as neon (a closed 2*p* shell) and is therefore diamagnetic. A substitution of Al(III) for Fe(III) reduces magnetic interactions very strongly,

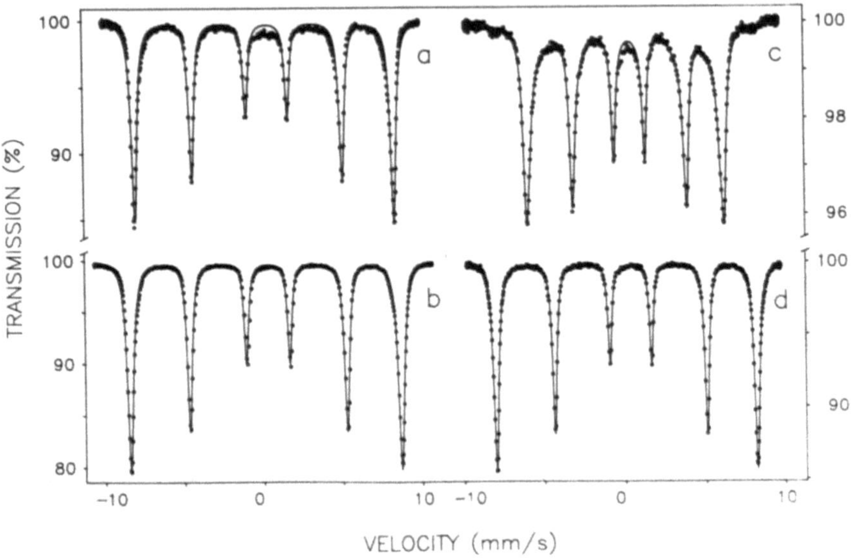

Figure 7. Mössbauer spectra of (a), (b) a hematite-rich ferricrete from South Africa and (c), (d) a hydrothermal goethite from Cornwall. (a), (c) At room temperature. (b), (d) At 4.2 K. For spectrum (a) only the sextet subspectrum ensuing from hematite is shown. In spectra (c) and (d) a minor hematite content has been ignored; spectrum (c) is fitted with three sextets to account for a distribution of magnetic hyperfine fields.

and thus has a marked influence on the magnetic properties of iron oxides. The influence of such substitutions on the Mössbauer spectra of iron oxides was first described in Janot et al. (1968) and has been used to determine the degree of aluminum substitution in hematite (Janot and Gibert, 1970). The magnetic hyperfine field of hematite, however, also varies as a function of sample crystallinity (Murad and Schwertmann, 1986), so that such an application is only justified for hematites that are essentially perfectly crystallized. A particle size reduction from 420 to 210 nm in the *c* direction, for example, reduces the magnetic hyperfine field as much as an Al substitution of 1%. The characterization of aluminous goethites using room-temperature Mössbauer spectra is further complicated by the fact that an Al/(Al + Fe) ratio of about 12% reduces the Néel temperature below 295 K (Fleisch et al., 1980). This means that goethites with an Al substitution of 12% or more are paramagnetic at room temperature. The substitution of elements other than aluminum in goethites, such as silicon (Murad, 1979a; Johnston and Norrish, 1981) or manganese (Vandenberghe et al., 1986), has also been indicated and may complicate an interpretation of hyperfine field data of this mineral further.

Room-temperature Mössbauer spectra even of apparently well-crystallized goethites with little or no foreign element substitution have asymmetric lines that are adverse to computer-fitting by Lorentzians. Such spectra can be fitted with distributions of Lorentzians (Wivel and Mørup, 1981), which can, in turn, be used to characterize the goethites (Murad, 1982a).

Room-temperature Mössbauer spectra of magnetite consist of two sextets: one each resulting from Fe(III) on the tetrahedral sites and from both Fe(II) and Fe(III) on the octahedral sites [Figure 8(a)]. Thermally activated electron delocalization, which takes place between the latter iron atoms, is fast enough for the nuclei to sense an average iron valence of 2.5, giving a correspondingly low hyperfine field of 46 T. Progressive oxidation of Fe(II) on the octahedral sites leads to a gradual disappearance of this spectral component. The Mössbauer spectrum of maghemite (completely oxidized magnetite) consists of a broadened sextet due to Fe(III) on the tetrahedral and octahedral sites [Figure 8(b)]. The individual contributions of these sites are difficult to resolve without the application of an external magnetic field.

Pedogenic iron oxides are often superparamagnetic due to their small particle size. Superparamagnetic breakdown of magnetic order has been reported for hematites smaller than 8 nm and for goethites smaller than 15 nm at room temperature (Janot et al., 1973). The Mössbauer spectra of these minerals will then resemble those of paramagnetic iron oxides (lepidocrocite), some dioc-

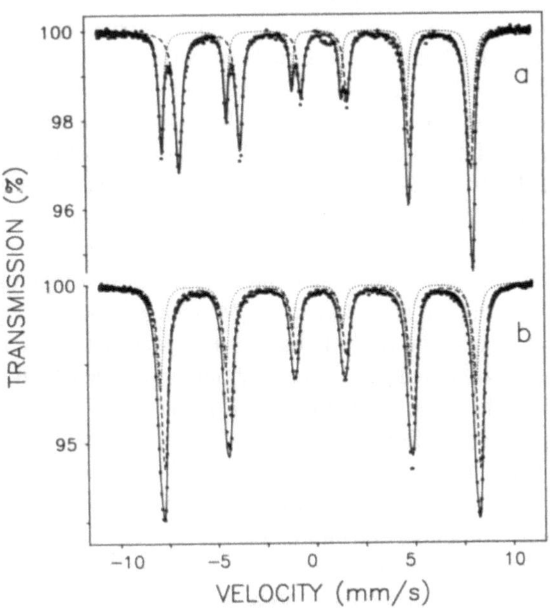

Figure 8. Room-temperature Mössbauer spectra. (a) Stoichiometric magnetite fitted with two sextets corresponding to (dotted subspectrum) tetrahedral Fe(III) and (broken subspectrum) octahedral Fe of nominal valence 2.5 (b) Maghemite fitted with two incompletely resolved sextets corresponding to tetrahedral and octahedral Fe(III). [Modified from (a) Murad and Johnston (1987) and (b) from Murad (1988).]

tahedral clay minerals, and diamagnetic iron sulfides, and are therefore of no diagnostic use. A special feature of superparamagnetic oxides is that the splitting of the doublet increases as particles become very small (Kündig et al., 1966; Kraan, 1973), so that the spectra can be fitted with two Lorentzian doublets instead of one. This behavior is similar to that shown by ferrihydrite at room temperature (Childs and Johnston, 1980; Murad and Schwertmann, 1980). The most likely explanation is that distributions of different atomic environments in very poorly crystallized iron oxide particles lead to the development of distributions of quadrupole-split doublets, which can be adequately approximated by two-doublet fits. It is fortuitous that paramagnetic akaganeite should have similar Mössbauer parameters, which has caused some confusion in the past (Childs and Johnston, 1980).

2. 77 K Spectra

There are two principal reasons for studying iron oxides by Mössbauer spectroscopy at low temperatures: magnetic order may be induced in some iron oxides that are inherently paramagnetic at room temperature, and superparamagnetic spin fluctuations of small particles may be slowed down sufficiently to restore magnetic splitting. While akaganeite (β-FeOOH) is the only pure iron oxide described here that is paramagnetic at room temperature but magnetically ordered at 77 K, such a temperature reduction also enables the characterization of poorly crystalline or highly aluminum-substituted goethites by Mössbauer spectroscopy (Murad and Bowen, 1987). On the other hand, the extent to which superparamagnetic fluctuations can be slowed down depends on the actual crystal size of the particles under consideration, and may require even lower temperatures than 77 K for iron oxides of extremely poor crystallinity.

Although there is no reliable X-ray diffraction evidence for the existence of akaganeite in soil environments, there have been repeated reports of the occurrence of this mineral in soils based on Mössbauer spectroscopic data. The identification of akaganeite by Mössbauer spectroscopy is commonly based on the fact that the *average* magnetic hyperfine field of akaganeite at 77 K (about 46 T) is lower than that of pure, bulk goethite. Hogg et al. (1975), however, showed that such lower hyperfine fields are much more likely to result from the presence of aluminous goethite than akaganeite. Careful Mössbauer spectroscopic work has furthermore shown that magnetically ordered akaganeite has at least three discrete magnetic components (Murad, 1979b; Childs et al., 1980), whereas in papers implying the presence of akaganeite in soils only single sextets were fitted to this component. There is at present no reason to assume that akaganeite should be a common mineral in soils, in particular since laboratory syntheses indicate the presence of either chlorine or fluorine to be essential for the formation of this mineral. Papers suggesting the contrary on the basis of Mössbauer spectra alone must be regarded with caution.

The possibility of observing magnetic order over the whole range of possible aluminum substitution stimulates the use of Mössbauer spectroscopy to determine aluminum substitution in goethites. Golden et al. (1979), however, showed that the crystallinity of goethites as reflected in their surface area can also affect the magnetic hyperfine field at 77 K. This limits the use of Mössbauer spectroscopy for this purpose to samples of essentially perfect crystallinity, since ignoring the influence of crystallinity may lead to too high aluminum substitutions being deduced from hyperfine field data (Ibanga et al., 1983).

This work was criticized by Fysh and Clark (1982a), who contended that the spectra of Golden et al. (1979) were in fact partially relaxed, but of too poor statistics for this relaxation to show up. Fysh and Clark (1982a) asserted that the influence of the surface area on the magnetic hyperfine field of goethite observed by Golden et al. (1979) should be absent at lower temperatures, and recommended determination of the aluminum substitution of goethites by Mössbauer spectroscopy at 4.2 K.

3. 4.2 K Spectra

At 4.2 K essentially all iron oxides and some iron-rich silicates, carbonates, sulfates, and so on, will be magnetically ordered. Magnetic hyperfine fields thus offer an additional possibility for the distinction of these minerals. On the other hand, Mössbauer spectra of mineralogically complex samples containing several constituents that are magnetically ordered at 4.2 K may have overlapping spectral components. This can complicate computer analysis considerably and may lead to ambiguous fits.

Mössbauer spectra of hematite- and goethite-containing soils (Schwertmann et al., 1982a) and samples from a hydrothermal assemblage (Murad, 1982b) have shown that both hematite and goethite may still be partly (super) paramagnetic near 77 K. Thus significantly lower temperatures may sometimes be required to induce complete magnetic order even in these minerals. At 4.2 K superparamagnetic spin fluctuations can, however, be expected to have come to a standstill.

Mössbauer spectra taken at 4.2 K can also be used to determine the influence of aluminum and particle size on the saturation hyperfine fields of very poorly crystallized and highly aluminous iron oxides without interference from relaxational effects. Based on the assumption that any influence of crystallinity on the magnetic hyperfine field of goethite must be due to superparamagnetic relaxation (and should therefore be absent at 4.2 K), Fysh and Clark (1982a) set up a relation between the magnetic hyperfine field of aluminous goethites at 4.2 K and the aluminum substitution. Murad and Schwertmann (1983), in contrast, showed that magnetic hyperfine field reductions in goethite at 4.2 K may result from both aluminum substitution and small particle size. This particle size dependence is an inherent property of small particles and bears no relation to relaxation. If the influence of particle size is ignored, too high aluminum substitutions will be deduced for poorly crystallized goethites even at 4.2 K.

The Morin transition in hematite (ideally at 260 K; see p. 138) is generally shifted to lower temperatures by both foreign element substitutions and a reduction of the particle size. Aluminum substitutions of $\geqslant 10$ mol % completely suppress the Morin transition even in perfectly crystallized hematites. Reductions of particle size to <20 nm in unsubstituted hematites have the same effect (Nininger and Schroeer, 1978; Schwertmann and Murad, 1983). The effects of particle size reduction, aluminum substitution, and possible substitutions by other elements add up, so that the observation of a Morin transition in soil hematites would be considered quite unusual (Bowen and Weed, 1984).

Most natural and synthetic ferrihydrites described so far require Mössbauer spectra at temperatures below 77 K if magnetic order is to be observed. No unique set of Mössbauer parameters can, however, be given for magnetically ordered ferrihydrites even at 4.2 K, which have to be fitted with distributions of magnetic hyperfine fields. The maxima of these distributions (46.5 to 50 T) and their half-widths (8 to 4.5 T) can be used to characterize the samples: poorer crystallinity leads to wider hyperfine field distributions and shifts the distributions as a whole to lower fields (Murad and Schwertmann, 1980; Murad, 1982c). These spectra are generally related to spectra in the paramagnetic state such that a reduction of crystallinity usually leads to a wider distribution of quadrupole splittings. In spite of the high quadrupole splitting in the paramagnetic state, magnetically ordered ferrihydrite has a small quadrupole interaction (between about −0.02 and −0.10 mm/s), which appears to vary depending on the randomization of the angle between the magnetic hyperfine field and the electric field gradient. The substitution of other elements for iron in ferrihydrite could possibly weaken magnetic interactions in the structure further and lead to an additional reduction of magnetic hyperfine fields (Childs et al., 1984). A lower quadrupole splitting in the superparamagnetic state, indicating a lower site distortion (i.e., an improved crystallinity), combined with a reduced hyperfine field (i.e., a weakening of magnetic interactions), has been observed for ferrihydrites that are intimately intergrown with other minerals, such as with silica and allophane in volcanic soils (Murad and Schwertmann, 1988a) or with manganese oxides in marine ferromanganese nodules (Murad and Schwertmann, 1988b).

The low Néel temperature of lepidocrocite (77 K) also requires measurements at temperatures significantly below 77 K if this is to be identified by Mössbauer spectroscopy. At 4.2 K lepidocrocite has a relatively low hyperfine field of about 46 T, which might lead to some confusion with ferrihydrite (especially since magnetically ordered lepidocrocite also has a small quadrupole interaction). Murad and Schwertmann (1984) and DeGrave et al. (1986), however, showed that the hyperfine fields even of poorly crystallized lepidocrocites are considerably better defined (i.e., these have significantly narrower hyperfine field distributions) than those of ferrihydrite.

B. Silicates

Magnetic order in silicates is described here less on its own account, but rather because the Mössbauer spectra of magnetically ordered silicates may resemble those of Fe(III) oxides. Iron-rich silicates have been reported to order magnetically at temperatures from 118 to below 4.2 K (Coey, 1984). With the exception of some more exotic minerals, magnetically ordered silicates can be expected to occur in iron-rich ultramafic rocks or in assemblages that contain iron-rich amphiboles, clay minerals, or garnets. It may therefore be necessary to exercise some care in interpreting low-temperature spectra of soils developed on parent rocks containing such minerals.

Magnetically ordered Fe(II) silicates can have complicated Mössbauer spectra comprising "forbidden" transitions, but the magnetic hyperfine fields are distinctly lower than those of Fe(III) oxides. Magnetically ordered Fe(III) silicates, on the other hand, have hyperfine fields that lie in the same range as those of iron oxides. This may lead to some confusion in low-temperature spectra.

The Mössbauer spectra of magnetically ordered silicates, as those of paramagnetic minerals, reflect local iron environments and valences, and may comprise one or several sets of associated lines. For Fe(III) silicates, development of one sextet is to be expected for every discrete iron site that is occupied. Some typical examples are cronstedtite, which has a Néel temperature of about 10 K and two sextets with hyperfine fields of 49.4 and 44.2 T at 4.2 K (Coey et al., 1981); andradite, which has a Néel temperature of 11.5 K and two sextets with similar hyperfine fields of 51.9 T but different quadrupole interactions at 4.2 K (Murad, 1984); and nontronite, which may be magnetically ordered with hyperfine fields of 50.5, 44.7, and 41.2 T at 4.2 K (Cardile et al., 1986).

Paramagnetic relaxation, which has been described for kaolinite (Fysh et al., 1983), leads to the development of Mössbauer spectra that resemble those of partially magnetically ordered minerals. Paramagnetic relaxation is probably more common than has been realized to date, but the statistics of spectra of the necessarily iron-poor minerals tend to be adverse to the recognition of this phenomenon.

V. Complex Natural Assemblages: Soils and Sediments

Mössbauer spectroscopy has attained an increasing popularity for the study of soils and sediments in recent years. Outstanding merits of Mössbauer spectroscopy are its sensitivity to the oxidation state of iron and to the presence of magnetically ordered phases. This is, however, neither a fast nor a cheap technique. While Mössbauer spectra of natural samples of complex iron mineralogy in the paramagnetic state can be deceptively simple (Hilton et al., 1986), comprehensive information on magnetically ordered phases may require several spectra of a sample to be taken at different temperatures. The results often cannot be placed in direct relation to structural effects (in contrast, for example, to X-ray diffraction), so that considerable expertise may be required to interpret spectra correctly. The use of Mössbauer spectroscopy is therefore not recommended as a *replacement* for standard techniques in soil mineralogy, but rather to complement results ensuing from such techniques or in cases where these fail (such as for the identification of iron oxides at low concentrations).

In this section, possibilities for the application of Mössbauer spectroscopy to the characterization of natural samples—which are usually complex, so that several minerals can contribute to the spectra—are outlined. Special emphasis is placed on mineral associations that are common in weathering environments, and examples are given where Mössbauer spectroscopy provides information that is difficult or impossible to obtain by other methods.

A. The Goethite–Hematite Pair

Well-crystallized, chemically pure hematite has a high magnetic ordering temperature of 955 K (as compared to 400 K for goethite). Natural hematites are often better crystallized and therefore less prone to superparamagnetic relaxation than coexisting goethites (Schwertmann, 1988), and hematites commonly accommodate only about half as much aluminum in their structures as goethites do (Schwertmann and Kämpf, 1984). These factors usually add up, so that natural hematites order magnetically at higher temperatures than goethites. Because of this and the high magnetic hyperfine field of hematite, room-temperature Mössbauer spectra are often useful for the identification and characterization of this mineral in samples of compex iron oxide mineralogy. In some soils the average particle size of hematites may nevertheless be so small that superparamagnetic relaxation causes the magnetic order to break down at room temperature. Even in such cases, however, the existence of *distributions* of particle size rather than discrete particle sizes may cause a minor proportion of the hematites to be large enough for the development of magnetic order, thus allowing a qualitative identification of this mineral.

In contrast to hematite, many synthetic and practically all pedogenic goethites require Mössbauer spectroscopy to be carried out at temperatures in the liquid nitrogen range to observe magnetic order. Very highly aluminum-substituted goethites, especially if they are of poor crystallinity, may have to be cooled down to temperatures significantly lower than 77 K to attain complete magnetic order (Goodman and Lewis, 1981; Murad and Bowen, 1987). As mentioned, such behavior can occasionally be observed for natural goethites and, in spite of the high Curie temperature, for hematites too.

Thus, even for a semiquantitative estimation of hematite/goethite ratios, Mössbauer spectra of soils will usually have to be taken at low tempeatures. Because hyperfine field distributions in goethites may furthermore persist to below 77 K, a quantification of hematite and goethite could require spectra taken at this temperature to be fitted with hyperfine field distributions rather than just one sextet for each component (Amarasiriwardena et. al., 1986). The hyperfine field of goethite approaches that of hematite more closely as temperatures are reduced, but the saturation fields (50.6 and 54.2 T for unsubstituted goethite and antiferromagnetic hematite, respectively) are still different enough to warrant a distinction of these minerals at 4.2 K, even when the spectral lines are broadened by hyperfine field distributions. If quantitative determinations of iron oxide mineral ratios are required, provision must be made for the dependence of recoil-free absorption on mineral type (Meisel and Kreysa, 1973) and temperature. Further corrections may be necessary if the iron oxides are aluminum substituted or poorly crystalline (De Grave et al., 1985; Murad, 1985).

The most frequent application of Mössbauer spectroscopy to soil science has probably been an assessment of goethite/hematite ratios. Bigham et al. (1978a) showed that goethite and hematite can be reliably identified in soils using Mössbauer spectra taken at low temperatures, in spite of the fact that the magnetic hyperfine fields of these minerals in soils tend to be lower than those of the chemically pure, well-crystallized species. A correlation of hematite/goethite ratios with soil color can also be observed using Mössbauer spectroscopy, the soils being redder the higher this ratio is (see Bigham et al., 1978b; Childs et al., 1978). Early reports on the possible occurrence of akaganeite in association with hematite in soils (Gangas et al., 1973) probably also relate to goethite–hematite assemblages (Hogg et al., 1975; see Section IV-A2.2).

Because Mössbauer spectroscopy can be used to identify magnetically ordered minerals—provided these have distinctive parameters—at very low concentrations, this technique is particularly well suited for the study of goethite and hematite in soils or other samples of complex mineralogy containing these minerals in concentrations that are too low to give adequately strong X-ray diffraction lines. Hematite has, for example, been identified in soils at a hematite/goethite ratio of 0.03 and a concentration of about 0.2%, that is, well below the detection limit of other instrumental techniques such as X-ray diffraction (Schwertmann et al., 1982a; Figure 9). Goethite and hematite have been identified by Mössbauer spectroscopy at even lower concentrations in limestones

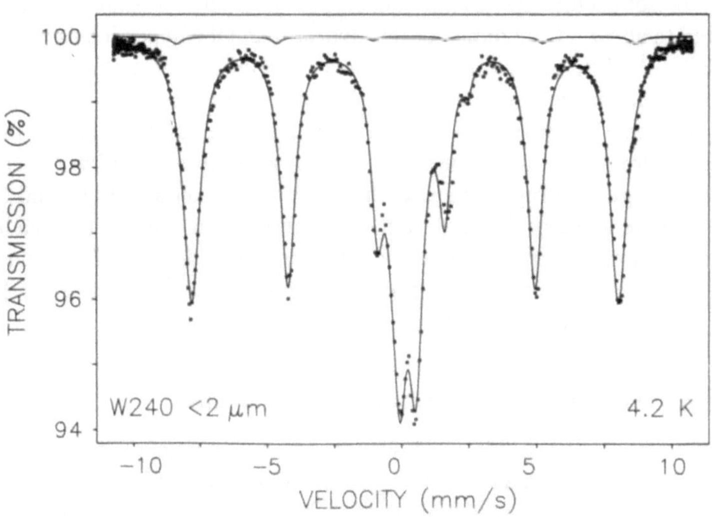

Figure 9. Mössbauer spectrum of the B_t (argillic horizon) horizon of a rubified Alfisol from the northern foreland of the Alps with a hematite/goethite ratio of 0.03 taken at 4.2 K. To emphasize the hematite contribution, only the subspectrum of this mineral has been plotted in addition to the envelope. [Modified from Schwertmann et al. (1982a).]

following calcite dissolution by a moderate acid treatment (Schwertmann et al., 1982a; Bronger et al., 1983).

B. The Goethite–Ferrihydrite Pair

In contrast to goethite and hematite, the distinction of goethite and ferrihydrite by Mössbauer spectroscopy *always* requires spectra to be taken below 77 K. Goethite may be partially or completely magnetically ordered between room temperature and 77 K, but ferrihydrite is usually superparamagnetic to below 77 K, preventing a distinction of Mössbauer spectra of ferrihydrite from those of superparamagnetic goethite or silicate-bound Fe(III). Mössbauer spectra taken at 77 K thus can be useful to characterize goethite–ferrihydrite associations provided goethite is magnetically ordered and the mineral assemblage contains no other minerals that can contribute to the Mössbauer spectra (Schwertmann et al., 1987), but as a rule spectra will have to be taken at significantly lower temperatures. At 4.2 K both goethite and ferrihydrite are magnetically ordered, but the hyperfine fields (about 50 T for goethite, and a distribution of fields with a maximum between about 46.5 and 50 T for ferrihydrite) of these minerals overlap to some extent. A deconvolution by computer fitting the contributions of goethite and ferrihydrite to Mössbauer spectra of composite samples containing both of these in the magnetically ordered state is not possible, so that a direct distinction of these minerals as for the goethite–hematite pair cannot be effected.

Schwertmann et al. (1982b) showed that two differences in the Mössbauer spectra of magnetically ordered goethite and ferrihydrite can help distinguish between these: goethite has a larger quadrupole interaction (usually −0.25 mm/s compared to about −0.05 mm/s for ferrihydrite), and the hyperfine field distributions of ferrihydrite are significantly wider than those that goethite may have. The differences in quadrupole interaction lead to the development of average values in composite samples. Such average quadrupole splittings can be used for diagnostic purposes provided the proportions of goethite and ferrihydrite are such that the average quadrupole interaction lies between about −0.10 and −0.20 mm/s. Upon removal of ferrihydrite by acid ammonium oxalate, the quadrupole interaction approaches that of pure goethite. Similarly, the hyperfine field distributions of samples containing both of these minerals will be significantly narrowed after selective dissolution of ferrihydrite (Figure 10).

C. Lepidocrocite–Goethite/Ferrihydrite Associations

As for the other Fe(III) oxides, Mössbauer spectra of paramagnetic lepidocrocite are not specific. Mössbauer spectra of lepidocrocite have to be taken at temperatures below 77 K if magnetic order is to be observed. Because of this, few Mössbauer spectroscopic data on natural lepidocrocite-containing assemblages are available to date. The magnetic ordering behavior of lepidocrocite is furthermore somewhat complicated by the coexistence of a paramagnetic and a magnetically ordered phase over a relatively wide temperature range of ≥10K (Johnson, 1969; Murad and Schwertmann, 1984).

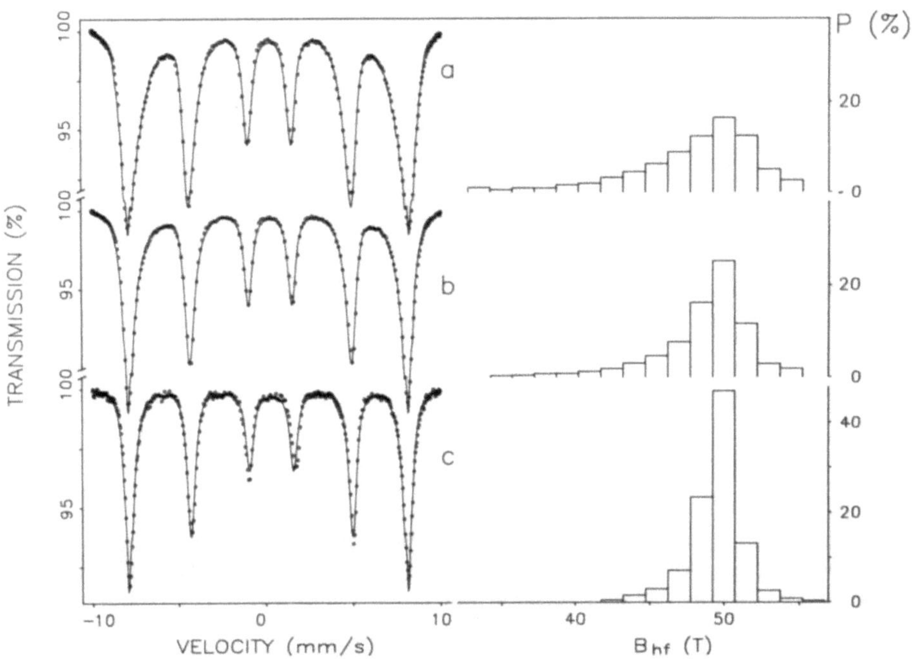

Figure 10. Mössbauer spectra of a concretion from a pipe stem containing goethite and ferrihydrite at 4.2 K and the corresponding distributions of hyperfine fields. (a) Before oxalate treatment. (b) Following three 2-hour treatments with ammonium oxalate. (c) Following a 24-hour oxalate treatment. [Modified from Schwertmann et al. (1982b).]

At 4.2 K lepidocrocite has a low quadrupole splitting of 0.02 mm/s and a relatively low magnetic hyperfine field of 46 T. The differences in quadrupole splitting and hyperfine field at 4.2 K make a distinction between lepidocrocite and goethite in samples of complex iron oxide mineralogy relatively straightforward (Figure 11). A distinction of lepidocrocite from ferrihydrite is much more difficult. The morphology of soil lepidocrocites (typically thin, highly serrated platelets) may cause a further reduction of magnetic hyperfine fields and lead to the development of hyperfine field distributions. The use of selective dissolution procedures holds little promise, because both lepidocrocite and ferrihydrite are soluble in both dithionite and oxalate. In samples containing only one of these minerals, a narrower distribution of magnetic hyperfine fields (half-width of distribution about 2 to 5 T) and a lower maximum of this distribution (at about 44.5 T) than for ferrihydrite may indicate the presence of lepidocrocite (Murad and Schwertmann, 1984; DeGrave et al., 1986).

Mössbauer spectra of a Mollisol from the Ha'apai Group, Tonga, showed completely paramagnetic behavior at 78 K, but predominant magnetic order at 4.2 K, about 85% of the spectral area consisting of one or several overlapping sextets with broad lines. This magnetic resonance could be fitted by two sextets with hyperfine fields of 44.5 and 48.9 T and quadrupole splittings of −0.01 and 0.08 mm/s, which were considered to result from aluminous lepidocrocite (the

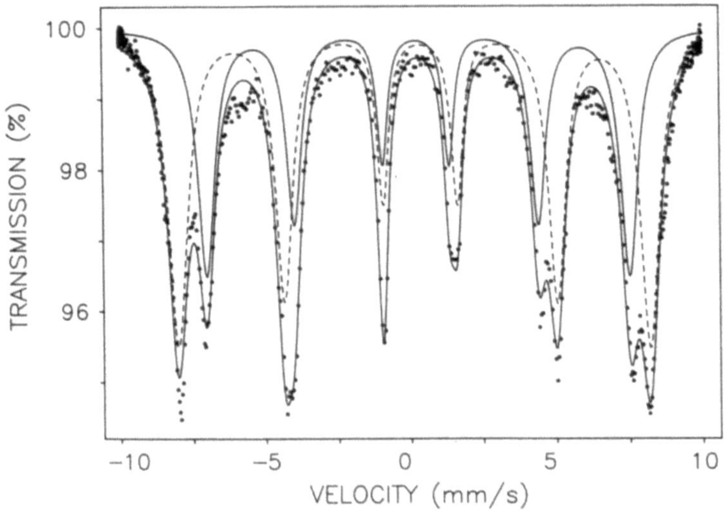

Figure 11. Mössbauer spectrum of an iron oxide sample consisting of goethite and lepidocrocite from the Siegerland ore province, Germany, at 4.2 K. The lepidocrocite subspectrum is indicated by a solid line, the goethite subspectrum by a broken line.

presence of which was also indicated by X-ray diffraction) and ferrihydrite, respectively (Childs and Wilson, 1983).

D. The Magnetite–Maghemite Pair

Identification of stoichiometric magnetite by Mössbauer spectroscopy is relatively straightforward: most soil magnetites are of detrital origin and therefore relatively well crystallized. Pure magnetite has a high Néel temperature of 850 K, and the room-temperature Mössbauer spectrum of magnetite ideally consists of two distinctly different sextets with hyperfine fields of 49 and 46 T. Oxidation of magnetite means that Fe(II) on the octahedral sites will be oxidized, as a result of which the latter sextet will disappear (see Section IV-A1). This renders deviations from magnetite stoichiometry quite evident in Mössbauer spectra, and rough estimates of the degree of nonstoichiometry have been made on natural magnetites by determining the ratio of the total Fe(III) to the Fe(II)+Fe(III) octahedral resonance (Papamarinopoulos et al., 1982).

Rather than being pure end members of the composition Fe_3O_4, natural magnetites are usually magnetite-near mixed crystals of the spinel series, in which different cations may substitute for both Fe(II) and Fe(III). The substitution of Ti(IV) for Fe(III), leading to mixed crystals of the composition $Fe_{3-x}Ti_xO_4$ (where $0 < x < 1$), is particularly common (and therefore well documented). The fact that such "titanomagnetites" may have suffered some oxidation can increase the number of parameters necessary for their characterization considerably, and thereby complicates their study in practice. In selected environments, such as weathering suites ensuing from the same parent material, the oxidation of magnetite can, however, be followed quite conveniently by room-temperature Mössbauer spectroscopy.

The contributions of the antiparallel spins in magnetite and maghemite split up under an externally applied magnetic field. The use of an external field thus facilitates the study of mixed crystals along the join magnetite–maghemite considerably, but it requires access to a superconducting magnet. Room-temperature spectra of such partially oxidized magnetites of the general composition $Fe_{3-x}O_4$ ($x < 0.33$) consist of three constituent patterns, which can be assigned to tetrahedral Fe(III), octahedral Fe(III), and a remaining contribution by iron of intermediate valence (Coey et al., 1971; Annersten and Hafner, 1973). The possiblity that other elements may substitute for iron can, however, also limit the value of such measurements unless data on the magnetite composition are available.

VI. Summary and Suggested Applications

Characteristic properties of minerals that can be monitored by Mössbauer spectroscopy are the valence and coordination of iron and, if the minerals will order magnetically, the magnetic ordering temperatures and hyperfine fields. Isomorphous substitutions and crystalline imperfections may lead to deviations from the behavior of pure end-member minerals that affect their Mössbauer spectra and can therefore be detected by this technique.

Some deviations from ideal behavior (such as superparamagnetism) can effect quite radical changes in the physical properties of minerals. Because such changes are not always immediately evident, they may lead to the development of misleadingly simple or ambiguous Mössbauer spectra. An unequivocal characterization of samples containing such minerals may require several Mössbauer spectra to be taken at different temperatures and possibly over different velocity ranges, and can therefore be a rather time-consuming task.

The application of Mössbauer spectroscopy to the study of paramagnetic minerals is often based on the premise that there is no interference from iron oxides with nonideal behavior. Such a premise, however, cannot be taken for granted in such complex systems as soils tend to be. Samples of complex iron mineralogy usually need to be studied by *combinations* of several techniques such as Mössbauer spectroscopy, X-ray diffraction, chemical analysis, and selective dissolution procedures, and the use of Mössbauer spectroscopy alone may lead to results that are prone to misinterpretation.

Because magnetic order not only gives information on magnetic hyperfine fields as an additional parameter, but also splits the constituent lines of Mössbauer spectra of minerals containing high-spin Fe(III) more strongly than the quadrupole interaction does, iron oxides are usually easier to characterize by Mössbauer spectroscopy than, for example, silicates, carbonates, sulfides. The interpretation of magnetic hyperfine field data—although suitable for mineral *identification*—suffers from the inherent weakness that several effects can be responsible for hyperfine field reductions (mainly small particle size and aluminum substitution). The sensitivity of ^{57}Fe Mössbauer spectroscopy for magnetically ordered iron oxides, nevertheless, is a more than adequate compensation for this shortcoming.

Acknowledgments

I am indebted to U. Wagner (Technical University of Munich, Garching) for placing the spectrum shown in Figure 2 at my disposal and to U. Schwertmann (Freising-Weihenstephan) for a helpful review of the manuscript. This work has been supported by the Deutsche Forschungsgemeinschaft.

References

Amarasiriwardena, D. D., E. DeGrave, L. H. Bowen, and S. B. Weed. 1986. Quantitative determination of aluminum-substituted goethite-hematite mixtures by Mössbauer spectroscopy. *Clays Clay Mineral.* 34:250–256.

Amigó, J. M., and J. P. Fortuné. 1981. Mineralogical appliation of the Mössbauer effect to the study of siderites from Batère (France); comparison with siderites of hydrothermal origin. *N. Jb. Miner. Mh.* 237–244.

Annersten, H., and S. S. Hafner. 1973. Vacancy distribution in synthetic spinels of the series Fe_3O_4–γ-Fe_2O_3. *Z. Krist.* 137:321–340.

Bancroft, G. M., A. G. Maddock, and R. G. Burns. 1967. Applications of the Mössbauer effect to silicate mineralogy —I. Iron silicates of known crystal structure. *Geochim. Cosmochim. Acta* 31:2219–2246.

Bigham, J. M., D. C. Golden, L. H. Bowen, S. W. Buol, and S. B. Weed. 1978a. Iron oxide mineralogy of well-drained Ultisols and Oxisols: I. Characterization of iron oxides in soil clays by Mössbauer spectroscopy, X-ray diffractometry, and selected chemical techniques. *Soil Sci. Soc. Amer. J.* 42:816–825.

Bigham, J. M., D. C. Golden, S. W. Buol, S. B. Weed, and L. H. Bowen. 1978b. Iron oxide mineralogy of well-drained Ultisols and Oxisols: II. Influence on color, surface area, and phosphate retention. *Soil Sci. Soc. Amer. J.* 42:825–830.

Bowen, L. H., and S. B. Weed. 1981. Mössbauer spectroscopic analysis of iron oxides in soil. pp. 247–261. In: J. G. Stevens and G. K. Shenoy (eds.), *Mössbauer spectroscopy and its chemical applications*. Amer. Chem. Soc., Washington, DC.

Bowen, L. H, and S. B. Weed. 1984. Mössbauer spectroscopy of soils and sediments. pp. 217–242. In: R. H. Herber (ed.), *Chemical Mössbauer spectroscopy*. Plenum, NY.

Bowen, L. H., S. B. Weed, and J. G. Stevens. 1969. Mössbauer study of micas and their potassium-depleted products. *Amer. Mineral.* 54:72–84

Bronger, A., J. Ensling, P. Gütlich, and H. Spiering. 1983. Rubification of *terrae rossae* in Slovakia: A Mössbauer effect study. *Clays Clay Mineral.* 31:269–276.

Cardile, C. M., J. H. Johnston, and D. P. E. Dickson. 1986. Magnetic ordering at 4.2 and 1.3 K in nontronites of different iron contents: A ^{57}Fe Mössbauer spectroscopic study. *Clays Clay Mineral.* 34:233–238.

Childs, C. W, and J. H. Johnston. 1980. Mössbauer spectra of proto-ferrihydrite at 77 K and 295 K, and a reappraisal of the possible presence of akaganéite in New Zealand soils. *Aust. J. Soil Res.* 18:245–250.

Childs, C. W., and J. H. Johnston. 1980. Mössbauer spectra of proto-ferrihydrite at 77 K and 295 K, and a reappraisal of the possible presence of akaganéite in New Zealand soils. *Aust. J. Soil Res.* 18:245–250.

Childs, C. W., and A. D. Wilson, 1983. Iron oxide minerals in soils of the Ha'apai Group, Kingdom of Tonga. *Aust. J. Soil Res.* 21:489–503.

Childs, C. W., B. A. Goodman, and G. J. Churchman. 1978. Application of Mössbauer spectroscopy to the study of iron oxides in some red and yellow/brown soil samples from New Zealand. pp. 555–565. In: M. M. Mortland and V. C. Farmer (eds.), *International clay conference 1978*. Elsevier, Amsterdam.

Childs, C. W., B. A. Goodman, E. Paterson, and F. W. D. Woodhams. 1980. The nature of iron in akaganéite. *Aust. J. Chem.* 33:15–26.

Childs, C. W., D. P. E. Dickson, B. A. Goodman, and D. G. Lewis. 1984. Moessbauer parameters for ferrihydrites at 4 K. *Aust. J. Soil Res.* 22:149–154.

Coey, J. M. D. 1980. Clay minerals and their transformations studied with nuclear techniques: The contribution of Mössbauer spectroscopy. *Atom. Energy Rev.* 18:73–124.

Coey, J. M. D. 1984. Mössbauer spectroscopy of silicate minerals. pp. 443–509. In: G. J. Long (ed.), *Mössbauer spectroscopy applied to inorganic chemistry*, vol. 1. Plenum, NY.

Coey, J. M. D., and P. W. Readman. 1973. New spin structure in an amorphous iron gel. *Nature* 246:476–478.

Coey, J. M. D., A. H. Morrish, and G. A. Sawatzky. 1971. A Mössbauer study of conduction in magnetite. *J. Physique* 32 C1:271–273.

Coey, J. M. D., O. Ballet, A. Moukarika, and J. L. Soubeyroux. 1981. Magnetic properties of sheet silicates: 1:1 layer minerals. *Phys. Chem. Mineral.* 7:141–148.

DeGrave, E., and R. Vochten. 1985. An ^{57}Fe Mössbauer study of ankerite. *Phys. Chem. Mineral.* 12:108–113.

DeGrave, E., A. E. Verbeeck, and D. G. Chambaere. 1985. Influence of small aluminum substitutions on the hematite lattice. *Phys. Lett.* 107A:181–184.

DeGrave, E., R. M. Persoons, D. G. Chambaere, R. E. Vandenberghe, and L. H. Bowen. 1986. An ^{57}Fe Mössbauer effect study of poorly crystalline γ-FeOOH. *Phys. Chem. Mineral.* 13:61–67.

Diamant, A., M. Pasternak, and A. Banin. 1982. Characterization of adsorbed iron in montmorillonite by Mössbauer spectroscopy. *Clays Clay Mineral.* 30:63–66.

Dickson, D. P. E., L. Heller-Kallai, and I. Rozenson. 1979. Mössbauer spectroscopic studies of iron in organic material from natural sedimentary environments. *Geochim. Cosmochim. Acta* 43:1449–1453.

Dyar, M. D. 1984. Precision and interlaboratory reproducibility of measurements of the Mössbauer effect in minerals. *Amer. Mineral.* 69:1127–1144.

Evans, B. J., R. G. Johnson, F. E. Senftle, C. Blaine Cecil, and F. Dulong. 1982. The ^{57}Fe Mössbauer parameters of pyrite and marcasite with different parameters. *Geochim. Cosmochim Acta* 46:761–775.

Eymery, J.-P., P. Moine, and A. LeRoy. 1978. Application de la spectrométrie Mössbauer á l'étude d'un cas de pseudomorphose de la sidérose par l'hématite. *Bull. Minéral.* 101:393–394.

Fleisch, J., R. Grimm, J. Grübler, and P. Gütlich. 1980. Determination of the aluminum content of natural and synthetic alumogoethites using Mössbauer spectroscopy. *J. Physique* 41 C1:169–170

Fysh, S. A., and P. E. Clark. 1982a. Aluminous goethite: A Mössbauer study. *Phys. Chem. Mineral.* 8:180–187.

Fysh, S. A., and P. E. Clark. 1982b. Aluminous hematite: A Mössbauer study. *Phys. Chem. Mineral.* 8:257–267.

Fysh, S. A., J. D. Cashion, and P. E. Clark. 1983. Mössbauer effect studies of iron in kaolin. I. Structural iron. *Clays Clay Mineral.* 31:285–292.

Gangas, N. H., A. Simopoulos, A. Kostikas, N. J. Yassoglou, and S. Filippakis. 1973. Mössbauer studies of small particles of iron oxides in soil. *Clays Clay Mineral.* 21:151–160.

Golden, D. C., L. H. Bowen, S. B. Weed, and J. M. Bigham. 1979. Mössbauer studies of synthetic and soil-occurring aluminum-substituted goethites. *Soil Sci. Soc. Amer. J.* 43:802–808.

Goodman, B. A. 1976. The effect of lattice substitutions on the derivation of quantitative site populations from the Mössbauer spectra of 2:1 layer lattice silicates. *J. Physique* 37 C6:819–823.

Goodman, B. A. 1980. Mössbauer spectroscopy. pp. 1–92. In: J. W. Stucki and W. L. Banwart (eds.), *Advanced chemical methods for soil and clay minerals research.* Reidel, Dordrecht, Boston.

Goodman, B. A., and D. G. Lewis. 1981. Mössbauer spectra of aluminous goethites (α-FeOOH). *J. Soil Sci.* 32:351–363.

Goodman, B. A., and M. J. Wilson. 1973. A study of the weathering of a biotite using the Mössbauer effect. *Mineral. Mag.* 39:448–454.

Goodman, B. A., J. D. Russell, A. R. Fraser, and F. W. D. Woodhams. 1976. A Mössbauer and I. R. spectroscopic study of the structure of nontronite. *Clays Clay Mineral.* 24:53–59.

Heller-Kallai, L., and I. Rozenson. 1980. Dehydroxylation of dioctahedral phyllosilicates. *Clays Clay Mineral.* 28:355–368.

Heller-Kallai, L., and I. Rozenson. 1981. The use of Mössbauer spectroscopy of iron in clay mineralogy. *Phys. Chem. Mineral.* 7:223–238.

Helsen, J. A., and B. A. Goodman. 1983. Characterization of iron(II)-and iron(III)-exchanged montmorillonite and hectorite using the Mössbauer effect. *Clay Mineral.* 18:117–125.

Hilton, J., G. J. Long, J. S. Chapman, and J. P. Lishman. 1986. Iron mineralogy in sediments. A Mössbauer study. *Geochim. Cosmochim. Acta* 50:2147–2151.

Hogg, C. S., P. J. Malden, and R. E. Meads. 1975. Identification of iron-containing impurities in natural kaolinites using the Mössbauer effect. *Mineral. Mag.* 40:89–96.

Hrynkiewicz, A. Z., and D. S. Kulgawczuk. 1963. Hyperfine structure of the 14.4 keV gamma line of ^{57}Fe in some iron oxy-hydroxides investigated with the Mössbauer effect. *Acta Phys. Polonica* 25:689–692.

Huggins, F. E., G. P. Huffman, D. A. Kosmack, and D. E. Lowenhaupt. 1980. Mossbauer detection of goethite (α-FeOOH) in coal and its potential as an indicator of coal oxidation. *Int. J. Coal Geol.* 1:75–81.

Ibanga, I. J., S. W. Buol, S. B. Weed, and L. H. Bowen. 1983. Iron oxides in petroferric materials. *Soil Sci. Soc. Amer. J.* 47:1240–1246.

Janot, C., and H. Gibert. 1970. Les constituants du fer dans certaines bauxites naturelles étudiées par effet Mössbauer. *Bull. Soc. Franç. Minéral. Crist.* 93:213–223.

Janot, C., M. Chabanel, and E. Herzog. 1968. Étude d'une limonite par effet Mössbauer. *Bull. Soc. Franç. Minéral. Crist.* 91:166–171.

Janot, C., H. Gibert, and C. Tobias. 1973. Caractérisation de kaolinites ferrifères par spectrométrie Mössbauer. *Bull. Soc. Franç. Minéral. Crist.* 96:281–291.

Janot, C., H. Gibert, X. de Gramont, and R. Biais. 1971. Étude des substitutions Al-Fe dans des roches latéritiques. *Bull. Soc. Franç. Minéral. Crist.* 94:367–380.

Johnson, C. E. 1969. Antiferromagnetism in γ FeOOH. A Mossbauer effect study. *J. Phys. C* 2:1996–2002.

Johnston, J. H., and C. M. Cardile. 1985. Iron sites in nontronite and the effect of interlayer cations from Mössbauer spectra. *Clays Clay Mineral.* 33:21–30.

Johnston, J. H., and K. Norrish. 1981. A ^{57}Fe Mössbauer spectroscopic study of a selection of Australian and other goethites. *Aust. J. Soil Res.* 19:231–237.

Kistner, O. C., and A. W. Sunyar. 1960. Evidence for quadrupole interaction of Fe^{57m}, and influence of chemical binding on nuclear gamma-ray energy. *Phys. Rev. Lett.* 4:412–415.

Kraan, A. M. van der. 1973. Mössbauer effect studies of surface ions of ultrafine α-Fe_2 O_3 particles. *Phys. Stat. Sol. a* 18:215–226.

Kündig, W., H. Bömmel, G. Constabaris, and R. H. Lindquist. 1966. Some properties of supported small α-Fe_2 O_3 particles determined with the Mössbauer effect. *Phys. Rev.* 142:327–333.

Meisel, W., and G. Kreysa. 1973. Relative Mössbauer-Konstanten von Eisenverbindungen zur quantitativen Analyse von Gemischen. *Z. Anorg. Allg. Chem.* 395:31–36.

Morrish, A. H. 1980. Morphology and physical properties of gamma iron oxide. pp. 171–197. In: H. C. Freyhardt (ed.), *Crystals 2: growth and properties*. Springer-Verlag, Berlin.

Mössbauer, R. M. 1958a. Kernresonanzabsorption von Gammastrahlung in Ir^{191}. *Naturwiss.* 45:538–539.

Mössbauer, R. M. 1958b. Kernresonanzfluoreszenz von Gammastrahlung in Ir^{191}. *Z. Phys.* 151:124–143.

Murad, E. 1979a. Mössbauer spectra of goethite: Evidence for structural imperfections. *Mineral. Mag.* 43:355–361.

Murad, E. 1979b. Mössbauer and X-ray data on β-FeOOH (akaganéite). *Clay Mineral.* 14:273–283.

Murad, E. 1982a. The characterization of goethite by Mössbauer spectroscopy. *Amer. Mineral.* 67:1007–1011.

Murad, E. 1982b. Iron oxide mineralogy of a hydrothermal assemblage on Santorini Island, Aegean Sea. *Mineral. Mag.* 67:89–93.

Murad, E. 1982c. Ferrihydrite deposits on an artesian fountain in lower Bavaria. *N. Jb. Mineral. Mh.* 45–56.

Murad, E. 1984. Magnetic ordering in andradite. *Amer. Mineral.* 69:722–724.

Murad, E. 1985. The influence of aluminium substitution on the absorption of gamma-rays in hematite. *Phys. Lett.* 111A:79–82.

Murad, E. 1987. Mössbauer spectra of nontronites: structural implications and characterization of associated iron oxides. *Z. Pflanzenernähr. Bodenk.* 150:279–285.

Murad, E. 1988. Properties and behavior of iron oxides as determined by Mössbauer spectroscopy. pp. 309–350. In: J. W. Stucki, B. A. Goodman, and U. Schwertmann (eds.) *Iron in soils and clay minerals*. Reidel, Dordrecht, Boston.

Murad, E., and L. H. Bowen. 1987. Magnetic ordering in Al-rich goethites: Influence of crystallinity. *Amer. Mineral.* 72:194–200.

Murad, E., and J. H. Johnston. 1987. Iron oxides and oxyhydroxides. pp. 507–582. In: G. J. Long (ed.), *Mössbauer spectroscopy applied to inorganic chemistry,* vol. 2. Plenum, NY.

Murad, E., and U. Schwertmann. 1980. The Mössbauer spectrum of ferrihydrite and its relations to those of other iron oxides. *Amer. Mineral.* 65:1044–1049.

Murad, E., and U. Schwertmann. 1983. The influence of aluminium substitution and crystallinity on the Mössbauer spectra of goethite. *Clay Mineral.* 18:301–312.

Murad, E., and U. Schwertmann. 1984. The influence of crystallinity on the Mössbauer spectrum of lepidocrocite. *Mineral. Mag.* 48:507–511.

Murad, E., and U. Schwertmann. 1986. Influence of Al substitution and crystal size on the room-temperature Mössbauer spectra of hematite. *Clays Clay Mineral.* 34:1–6.

Murad, E., and U. Schwertmann. 1988a. The characterization of poorly crystalline Si-containing natural iron oxides by Mössbauer spectroscopy. *Hyperfine Interact.* 41:835–838.

Murad, E., and U. Schwertmann. 1988b. Iron oxide mineralogy of some deep-sea ferromanganese crusts. *Amer. Mineral.* 73:1395–1400.

Murad, E., and R. M. Taylor. 1984. The Mössbauer spectra of hydroxycarbonate green rusts. *Clay Mineral.* 19:77–83.

Murad, E., and R. M. Taylor. 1986. The oxidation of hydroxycarbonate green rusts. pp. 585–593. In: G. J. Long and J. G. Stevens (eds.), *Industrial applications of the Mössbauer effect.* Plenum, NY.

Nininger, R. A., and D. Schroeer. 1978. Mössbauer studies on the Morin transition in bulk and microcrystalline α-Fe_2O_3. *J. Phys. Chem. Solids* 39:137–144.

Papamarinopoulos, P., P. W. Readman, Y. Maniatis, and A. Simopoulos. 1982. Magnetic characterization and Mössbauer spectroscopy of magnetic concentrates from Greek lake sediments. *Earth Planet. Sci. Lett.* 57:173–181.

Pollak, H., and J. G. Stevens. 1986. Phyllosilicates: a Mössbauer evaluation. *Hyperfine Interact.* 29:1153–1156.

Rozenson, I., and L. Heller-Kallai. 1976a. Reduction and oxidation of Fe^{3+} in dioctahedral smectites—1: Reduction with hydrazine and dithionite. *Clays Clay Mineral.* 24:271–282.

Rozenson, I., and L. Heller-Kallai. 1976b. Reduction and oxidation of Fe^{3+} in dioctahedral smectites—2: Reduction with sodium sulphide solutions. *Clays Clay Mineral.* 24:283–288.

Rozenson, I., and L. Heller-Kallai. 1978. Reduction and oxidation of Fe^{3+} in dioctahedral smectites—3: Oxidation of octahedral iron in montmorillonite. *Clays Clay Mineral.* 26:88–92.

Rozenson, I., E. R. Bauminger, and L. Heller-Kallai. 1979. Mössbauer spectra of iron in 1:1 phyllosilicates. *Amer. Mineral.* 64:893–901.

Schwertmann, U. 1985. The effect of pedogenic environment on iron oxides. *Adv. Soil Sci.* 1:171–200.

Schwertmann, U. 1988. Some properties of soil and synthetic iron oxides. pp. 203–250. In: J. W. Stucki, B. A. Goodman, and U. Schwertmann (eds.), *Iron in soils and clay minerals.* D. Reidel. Dordrecht, Boston.

Schwertmann, U., and N. Kämpf. 1984. Properties of goethite and hematite in kaolinitic soils of southern and central Brazil. *Soil Sci.* 139:344–350.

Schwertmann, U. and E. Murad. 1983. Effect of pH on the formation of goethite and hematite from ferrihydrite. *Clays Clay Mineral.* 31:277–284.

Schwertmann, U., E. Murad, and D. G. Schulze. 1982a. Is there Holocene reddening (hematite formation) in soils of axeric temperate areas? *Geoderma* 27:209–223.

Schwertmann, U., D. G. Schulze, and E. Murad. 1982b. Identification of ferrihydrite in soils by dissolution kinetics, differential X-ray diffraction, and Mössbauer spectroscopy. *Soil Sci. Soc. Amer. J.* 46:869–875.

Schwertmann, U., L. Carlson, and E. Murad. 1987. Properties of iron oxides in two Finnish lakes in relation to the environment of their formation. *Clays Clay Mineral.* 35:297–304.

Simopoulos, A., A. Kostikas, I., Sigalas, N. H. Gangas, and A. Moukarika. 1975. Mössbauer study of transformations induced in clay by firing. *Clays Clay Mineral.* 23:393–399.

Stevens, J. G. 1983. Isomer shift reference scales. *Hyperfine Interact.* 13:221–236.

Taylor, R. M. 1980. Formation and properties of Fe(II)Fe(III) hydroxy-carbonate and its possible significance in soil formation. *Clay Mineral.* 15:369–382.

Taylor, R. M., and R. M. McKenzie. 1980. The influence of aluminum on iron oxides. VI. The formation of Fe(II)-Al(III) hydroxy-chlorides, -sulfates, and -carbonates as new members of the pyroaurite group and their significance in soils. *Clays Clay Mineral.* 28:179–187.

Vandenberghe, R. E., A. E. Verbeeck, E. DeGrave, and W. Stiers. 1986. ^{57}Fe Mössbauer effect study of Mn-substituted goethite and hematite. *Hyperfine Interact.* 29:1157–1160.

Vértes, A., K. Jónás, I. Czakó-Nagy, and E. Nemecz. 1981. A combined application of Mössbauer spectroscopy and chemical transformations for chemical analysis. *Radiochem. Radioanal. Lett.* 48:93-100.

Weaver, C. E., J. M. Wampler, and T. E. Pecuil. 1967. Mössbauer analysis of iron in clay minerals. *Science* 156:504–508.

Wivel, C., and S. Mørup. 1981. Improved computational procedure for evaluation of overlapping hyperfine parameter distributions in Mössbauer spectra. *J. Phys. E* 14:605–610.

Yassoglou, N. J., C. Nobeli, A. J. Kostikas, and A. C. Simopoulos. 1972. Weathering of mica flakes in two soils in northern Greece evaluated by Mössbauer and conventional techniques. *Soil Sci. Soc. Amer. J.* 36:520–527.

High-Resolution Transmission Electron Microscopy (HRTEM) in the Study of Clays and Soils

P.B. Malla* and S. Komarneni*†

I. Introduction

This review is intended for those soil scientists who have little or no previous knowledge of transmission electron microscopy (TEM). Our objective here is to introduce the usefulness and the limitations of high-resolution TEM(HRTEM) in the studies of soils and clays. Information is presented as to how far HRTEM (with accelerating voltages of 100 to 200 kV) can be extended to gather useful pieces of structural, physical, and chemical information of soil minerals, especially layer silicates, which are not obtainable by other techniques. We have therefore not attempted to present an exhaustive account of the theory and

*Materials Research Laboratory, Pennsylvania State University, University Park, PA 16802, U.S.A.

†Also with the Department of Agronomy.

principles of TEM. For such details, readers are referred to Hirsch et al. (1965 and Wenk (1978).

In the beginning readers will be introduced to some important differences between TEM and X-ray diffraction (XRD), followed by the principles of TEM and some commonly used specimen preparation techniques. The characterization section is divided into four subsections, which deal with the applications of HRTEM and the types of information that can be obtained from it. The references quoted are representative rather than exhaustive.

Although TEM was developed some 55 years ago, it did not emerge as a powerful tool for mineral studies until the early 1970s. Recent developments in HRTEM permit the direct observation of visual images of structural details of mineral samples, thereby providing useful information on defects, transformations, and physical and chemical characteristics on a nanometer or angstrom scale (Buseck and Iijima, 1974; Iijima and Buseck, 1978; Tessier, 1984; Vali and Koster, 1986; Amouric, 1987; Malla et al., 1987). Iijima and Allpress (1974), who were among the first to use HRTEM for direct structural observation, clearly demonstrated the power of HRTEM.

Soil is a composite of both organic and mineral fractions, which may be crystalline or amorphous in nature, or both. Although morphological information of fine-grained amorphous materials can be obtained by TEM, lattice fringe imaging of these materials to gain additional information is much more difficult because high-resolution imaging is associated with the diffraction of electrons. Diffraction of electrons requires that the atoms in the structure of the samples be distributed in a periodic fashion, that is, that the samples be crystalline (or partially crystalline) in nature. Therefore, individual or closely related soil minerals (or soil-forming minerals) should be separated from the bulk soil prior to TEM examination in order to get meaningful information on them. In this review high-resolution imaging of clay minerals, which are very important constituents of soils and sediments, is reviewed. The clay mineral composition of soils controls their agricultural, environmental, commercial, and other uses. Clay minerals are also important in petroleum generation and recovery and in numerous other industrial applications. They also serve as excellent indicators of the past history of the earth. Thus it is important to gain useful information on this group of minerals with as many techniques as possible, including HRTEM.

II. Transmission Electron Microscopy versus X-Ray Diffraction

X-ray diffraction techniques have been used successfully for the structural analysis of minerals for a long period of time. The X-ray techniques have also proved invaluable for routine identification of minerals. Both X-ray and electron microscopy methods are based on the principles of diffraction of waves. However, they differ in the nature of their interactions with substances, which leads to their specific applications.

1. Nature of Interactions of Electrons and X-Rays with Mass

Electrons are scattered by the electric potential of both the nucleus and the orbital electrons, whereas X-rays are diffracted only by orbital electrons. The difference in the scattering power of light and heavy atoms is more important in the case of electron diffraction as compared to X-ray diffraction. Therefore light atoms may be located more easily in the presence of heavy atoms. Electrons interact with atoms many times more strongly than X-rays, which leads to the phenomenon of multiple scattering (dynamic diffraction) as compared to single scattering (kinematic diffraction—reasonable assumption) in the case of X-rays. Since the dynamic interaction of electrons with matter affects the electron diffraction intensities, the intensities of electron diffraction, unlike X-ray intensities, cannot be used in a quantitative way. Kinematic approximation may, however, be done in cases where the thickness of the sample is very small (up to 150 Å).

2. Wavelength λ

For a transmission electron microscope with 100-kV accelerating voltage, the wavelength λ is 0.037 Å, which is almost two orders of magnitude smaller than the wavelength of X rays (for Cu K_α, $\lambda = 1.54$ Å). This leads to a much smaller Bragg angle Θ for electrons and an increased number of reflection orders for a family of planes that simultaneously fulfill the Bragg condition. Since electrons interact with matter much more strongly and have smaller wavelengths than X rays, electron diffraction shifts the boundary of the detectable crystallite size to a lower value. In other words, a material that is amorphous for X rays may be crystalline for electrons. Thus electron diffraction leads to a better understanding of the nature of our samples.

3. Size of Specimen

Theoretically, electron diffraction can provide information from volumes smaller than 10 Å on a side and 150 Å thick (volume $<10^4$ Å^3) whereas XRD requires that the sample be at least 0.1 mm on a side (volume $>10^{18}$ Å^3) (Buseck and Iijama, 1974). As a matter of fact, a specimen for TEM examination should be penetrable to electrons, and therefore the thickness of a good TEM specimen depends on its absorption which is a function of the atomic number and crystallographic orientation (usually <500 Å). In XRD, information obtained is averaged over many unit cells involving numerous particles (for example, 5 million particles), whereas in TEM detailed information can be obtained from a very small area (a 50 to 100-Å particle). One therefore should be cautious in interpreting TEM data taken from a small area only, since they might not be representative of the entire sample. This problem can be avoided by gathering a large number of data points at different places in several specimens.

4. Specimen Environment

Specimens in TEM are subjected to high vacuum, which causes dehydration and may alter their properties. The specimen may also be damaged by irradiation with electrons as a result of contamination, temperature rise, or breaking of the chemical bonds. Therefore, to prevent damage, specimens for TEM examination have to be prepared with appropriate care, and sometimes they have to be handled carefully during focusing in the TEM. These problems are seldom encountered in XRD.

5. Type of Information

From TEM, both imaging and diffraction information can be obtained on the same area by adjusting the projective lens system either on the focal plane (diffraction) or on the image plane (imaging). Images should bear a 1 : 1 relation to the structural information of the specimen, such as crystal lattice, defects, dislocations, and crystal boundaries. The image observed by HRTEM is the Fourier transform of the diffraction pattern, which is itself the Fourier transform of the direct lattice. It should, however, be realized that an image formed by the Fourier transform of the diffraction pattern may not necessarily be a straightforward representation of the structure of the sample, since multiple scattering of the electrons (dynamic interaction) usually occurs in the sample. Images of samples thicker than 150 Å may also be misinterpreted due to the change in contrast in lattice fringes (Eggleton, 1988) induced by multiple scattering.

6. Resolution

A TEM with 100 to 200-kV accelerating voltage can have a point-to-point resolution down to approximately 2.0 Å. Such a high resolution permits obtaining information on one- and two-dimensional defects, point defects, long-range ordering, twinning, and other phenomena occurring on the unit-cell level.

III. Principles of TEM

A. Layout of the Instrument

The transmission electron microscope is similar to most optical microscopes, but arranged in a reverse fashion. A typical TEM has an electron gun as the electron source, which is a heated tungsten filament at a high negative potential, typically 20 to 200 kV. The electrons emitted by the filament are accelerated by high voltage in vacuum and are brought to focus through the sample by a series of electromagnetic lenses. First in this series is the condenser lense assembly with field-limiting apertures that focus the electrons onto the sample. The scattered or transmitted electrons are focused with an objective lens lying below the sample, and a projector lens system gives a highly magnified image or diffraction pattern onto the viewing (fluorescent) screen or photographic plate. The layout of a TEM is given in Figure 1. A schematic representation of an ideal imaging system is presented in Figure 2. Here the back focal plane is the plane where the parallel beams from the specimen are brought to focus and the image plane is where all the rays from the sample converge to form a two-dimensional inverted image. In the back focal plane a diffraction pattern of the specimen is formed.

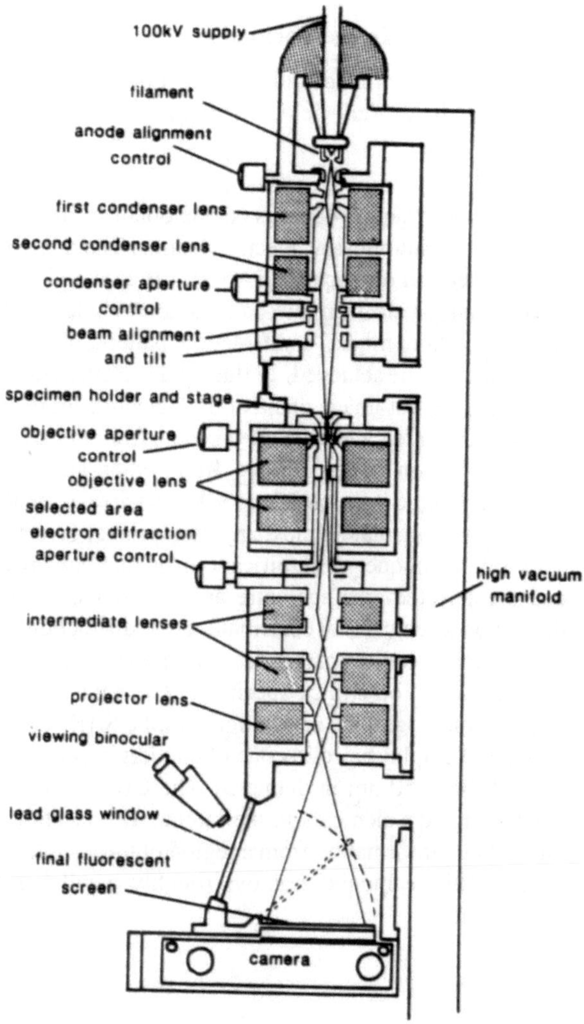

Figure 1. Layout of a TEM [From Nadeau and Tait (1987)].

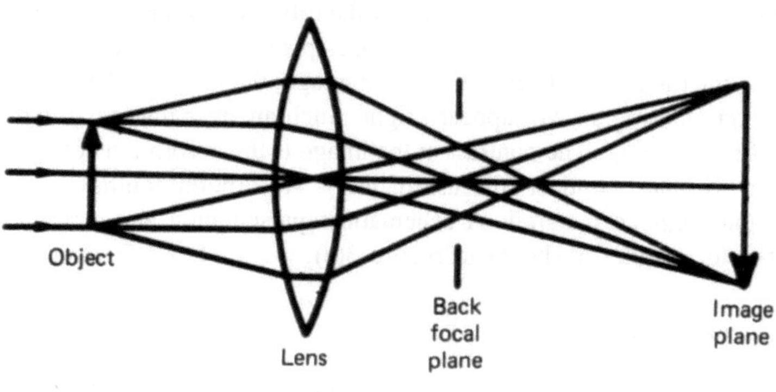

Figure 2. Schematic representation of ideal imaging system.

B. Diffraction Pattern and Image Formation

The TEM can be operated to produce either diffraction patterns or one of many types of images (bright-field, dark-field, or high-resolution images).

1. Diffraction Pattern

Many different types of diffraction patterns can be obtained from TEM, depending on the geometry of the illuminating electron beams. The three important ones are (1) selected-area electron diffraction (SAED), (2) convergent-beam electron diffraction (CBED), and (3) electron microdiffraction (Veblen, 1985). By using the selected-area diffraction mode, a diffraction pattern of a specific region selected by the operator can be achieved. In this mode an aperature (selected-area aperature) is placed in the first image plane of the objective lense coplanar to the image such that the transmitted and diffracted beams from the area selected only contribute to the diffraction pattern. Rays from other areas are essentially intercepted. By this technique it is possible to select an area as small as 0.15 μm, and therefore allows one to isolate almost any soil mineral crystal, which is not possible by the X-ray technique. A significant error may however occur if the image plane and the selected-area aperature are not coplanar because the rays from outside the region selected may contribute to the diffraction pattern. SAED is useful for determining lattice parameter and Laue group symmetry. The *Kikuchi pattern* is an SAED on a thick specimen. The Kikuchi lines are used to determine the exact crystallographic orientation. Unlike SAED, in which parallel beams are used, CBED employs a solid cone of converging electrons focused on a very small area of 20 to 50 nm in diameter. The CBED can be used to give information about symmetry elements in the lattice. The electron microdiffraction is also used to obtain information from a region of the sample as small as 4 to 10 nm in diameter. This technique employs roughly parallel or slightly convergent stationary beams.

2. Image Types

a. Bright-Field Image

A crystal in Bragg conditions will diffract the incident beam while still transmitting a part of it. In this situation, if all of the diffracted beams are intercepted by placing a small aperture in the back focal plane of the objective lens while allowing the transmitted beams to form an image, the image will appear dark and the background region will appear bright. Such an arrangment is referred to as bright-field imaging. The contrast in the image is the absorption contrast type as opposed to phase contrast (discussed later). Absorption contrast is typical of amorphous organic material. A schematic representation and a real image of bright-field imaging is shown in Figure 3(a).

b. Dark Field Image

In dark-field imaging an objective aperture is positioned to allow only one selected diffracted beam to form the image while intercepting all other diffracted

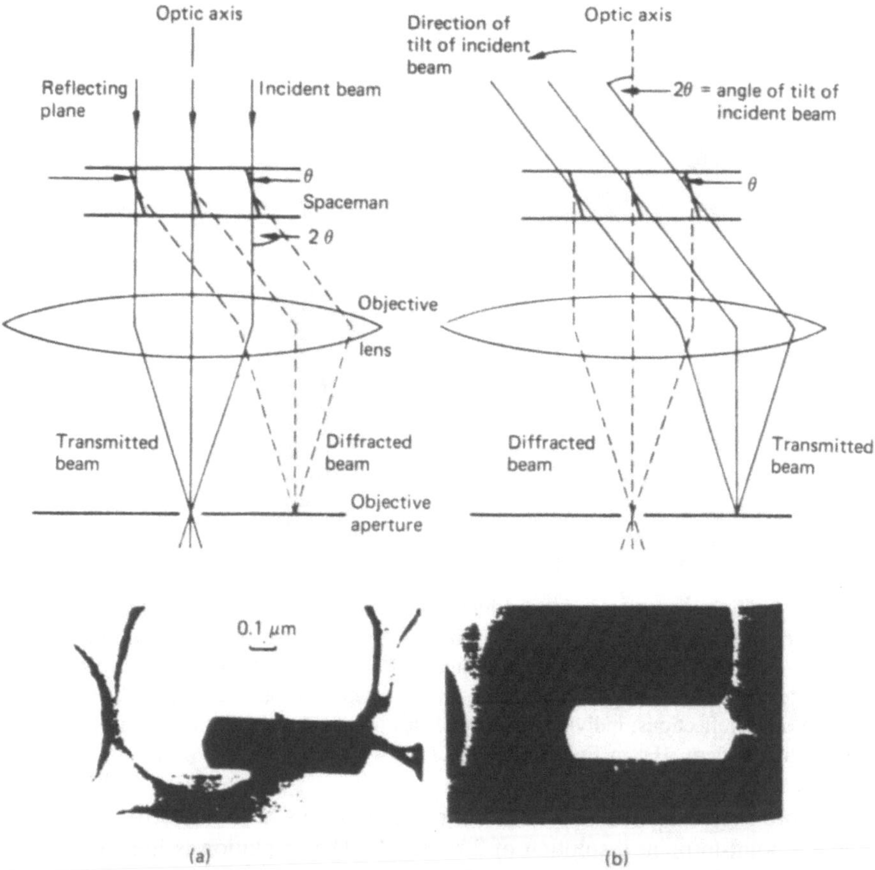

Figure 3. (a) Bright-field imaging schematic and real image. (b) Dark-field imaging schematic and real image. Images were formed using (a) zero-order beam and (b) 040 reflection of a goethite crystal. [From Eggleton (1988).]

and transmitted beams. In this case the image formed by diffracted beams appears bright, but the field of view (nondiffracting region) remains dark. This situation is shown in Figure 3(b).

c. High-Resolution (Phase-Contrast) Image

High-resolution images are achieved when one or several of the diffracted beams are allowed to pass through the objective aperture and are recombined with transmitted electrons, resulting in a phase interference between these beams. When one or two diffracted beams are recombined with transmitted beams (two- or three-beam images), one obtains a one-dimensional periodic fringe image resolving the individual lattice plane. On the other hand, if several diffracted beams are recombined with transmitted beams, two-dimensional images (multi-beam-structure images) are produced from an array of diffraction spots, and

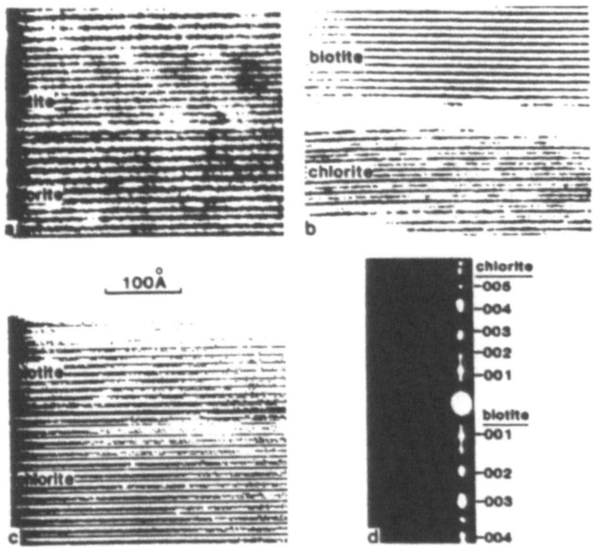

Figure 4. High-resolution images of biotite–chlorite intergrowths showing the effect of increased resolution due to increased size of the objective aperture. (a) First- and second-order 001 reflections; only 10-Å and 14-Å layer spacings are distinguishable. (b) First- to third-order reflections; brucite sheets in chlorite are resolved. (c) First- to fourth-order reflections; individual metals planes in biotite and chlorite are resolved. (d) diffraction pattern. [From Eggleton (1988).]

therefore more detailed structural information is obtained. Modern TEMs can give a point-to-point resolution of 2 to 3.0 Å. The resolution is limited by the diameter of the objective aperture. Figure 4 shows increased resolution due to increased aperture size (Eggleton, 1988). Further details on the theory of high-resolution imaging can be found in Cowley and Iijima (1976) and Edington (1975). It must, however, be realized that a one-to-one correspondence of the lattice plane or structure to the image formed depends, among other factors, on the various aberrations present in the lenses. Therefore one should take great care in the interpretation of such images.

IV. Sample Preparation

Specimen preparation is one of the most important and sometimes difficult and tedious parts of microstructural study by TEM because the resolution and the contrast in the electron image partly depend on how the specimen has been prepared. Since the specimen for TEM examination should be penetrable to electrons and electrons do not penetrate crystals as easily as X-rays, the thickness of the specimen is of paramount importance in TEM. The thickness of the specimen depends, however, on the type of sample and the type of information sought. For example, a sample as thick as 500 Å (50 nm) may be good enough for one-dimensional lattice fringe imaging of minerals, but a sample <150 Å (15 nm) is required to obtain good structural information (Buseck and Iijima, 1974; Eggleton, 1988). Therefore, before an attempt is made to examine a particular sample by HRTEM, it should be borne in mind whether a thin enough specimen can be prepared. It is also possible to damage or alter wholly or partly the internal structure of the sample by some of the sample preparation techniques. Great care should be taken while preparing a specimen, and a suitable technique for a specific purpose should be selected. There are several methods for specimen preparation for TEM examination: (1) particle dispersion, (2) fragmentation and cleavage, (3) ion thinning, (4) ultramicrotomy, (5) electropolishing, (6) chemical dissolution, and (7) replicas. In this chapter the three main methods, particle dispersion, ion thinning, and ultramicrotomy, which are widely used in high-resolution imaging, are briefly discussed. For more details, readers are advised to consult McKee and Brown (1977), Paulus et al. (1975), Comer (1971), Smart and Tovey (1982), Agar et al. (1978), and Spurr (1969).

A. Particle Dispersion

In this method, soil or clay samples should be fractionated to <2.0-μm particle diameter and dispersed thoroughly with an ultrasonic vibrator in alcohol or acetone. A small drop of suspended clay is deposited on a coated support film of carbon (holey carbon film) covering the grid and allowed to dry. The concentration of the suspended sample need to be controlled depending on the particle size, the type of sample, and the information desired. A long exposure of the sample to an ultrasonic disperser with very high frequency could destroy or cause defects in the sample. Another disadvantage of this method is that the observation can be made only in a plane defined by sedimentation.

B. Ion Thinning or Ion Milling

This method involves the bombardment of the sample with ions, preferably argon, until perforation or until it is thin enough to be examined by TEM. First sections (about 30 μm thick) are made by grinding and polishing or chemical etching. They may be glued to the electron microscope grid. The grid (specimen) is then placed in an ion mill where the sample is bombarded with 2- to 10-kV argon ions. This is a relatively reliable technique and produces less artifacts than most other techniques (Paulus et al., 1975). Various factors, such as the accelerating voltage of the ion, the type of sputtering ion (to obtain a high yield without deep penetration), and the grazing incidence (to prevent selective etching) should be controlled, and a multibeam cathode (to obtain uniform sputtering) should be used to obtain good defect-free specimens for TEM observation. One should, however, be advised that the soft materials are thinned preferentially to the hard ones; therefore this technique is not recommended for organomineral specimens such as bulk soil.

C. Ultramicrotomy

Ultramicrotomy is widely used in the preparation of biological samples, and recently it has also been proven invaluable in the study of minerals. With this method the specimen (crystal or slurry of fine powder) is embedded in a low-viscosity epoxy resin (Spurr, 1969) which is subsequently hardened. Impregnation reinforces the structure of the specimen and prevents quick damage of the sample under the electron beam. To study the stacking disorder and layer organization of clay minerals, it is necessary to examine the crystal structure in a direction parallel to the layers. Therefore layers have to be oriented parallel to each other for best results. A parallel orientation can be achieved by impregnating the sample under vacuum (Lee et al., 1975) or under known pressure (Tessier, 1984; Malla et al., 1987; a detail of the technique will be communicated elsewhere). The latter is especially useful to study the effect of applied pressure and hydration on the organization of layers in tactoids or quasicrystals. It is absolutely necessary to remove all water from the sample sequentially by replacing it with solvents, such as ethanol or methanol, and propylene oxide before embedding. The presence of any trace of water severely hinders the microslicing process. The impregnated and hardened sample is shaped to a pyramid by a glass knife and then sliced to a very thin section (500 Å or less) by a diamond knife (ultramicrotome). The thin section, which floats on water in the microtome device, is fished with a carbon-coated TEM grid. The technique is also used for the study of bulk soil, but the sectioning of hard and large-size minerals present in soil may distort the soil fabric and mineral morphology.

V. Characterization

HRTEM allows direct observation of the structure and the various types of defects present in the structure on a unit-cell level of the mineral. Consequently it provides useful information on their transformation, interstratification, and physical and chemical properties. The HRTEM provides more detailed information, which is not obtainable by other techniques, and it also complements other techniques greatly. In this section we discuss the types of information on mineral samples that can be gathered by HRTEM and present some examples.

A. Structural Features

Recent developments in HRTEM have transformed TEM into a powerful tool for one- and two-dimensional imaging of crystal structures (Allpress and Sanders, 1973). Iijima and Allpress (1974) first performed the direct structure determination of niobium–tungsten oxide ($4Nb_2O_5{\cdot}9WO_3$) that demonstrated the real power of HRTEM. They not only confirmed the structure of this oxide determined by X-ray technique, but they were also able to detect more details on empty and partially filled tunnels and fault boundaries. Such detailed and interpretable results on two-dimensional images of soil minerals are somewhat more difficult to obtain due to the effect of lower contrast and overlapping structural features. However, limited reports of structure images of layer silicates are available in the literature and are giving a great deal of insight into their defects, which leads to a better understanding of polytypism, growth habits, transformation, and so on (Yoshida and Suito, 1972; Buseck and Iijima, 1974; Iijima and Buseck, 1978; Buseck and Veblen, 1981; Buseck and Cowley, 1983; Spinnler et al., 1984; Amouric and Parron, 1985; Amouric, 1987).

As early as 1968, Brown and Rich (1968) were able to observe one-dimensional lattice images of a muscovite and altered muscovites. These images showed 10-Å periodicities, interstratification, possible stacking faults, dislocation or growth-step planes, and frayed edges. Yoshida and Suito (1972) successfully performed the lattice fringe imaging of montmorillonite expanded with organic molecules such as octadecylammonium ions. They were able to observe the interlayer distances d(001) of the clay in accordance with the size and orientation of the intercalated organic ions. They were also able to recognize local variations in expansion within a particular layer, which could well be due to the asymmetric charge distribution in the layer.

Buseck and Iijima (1974) succeeded in obtaining the crystal structure images of different types of silicate minerals. They were able to differentiate regular from perturbed regions containing defects and they were also able to observe antiphase domains and stacking faults. A high-magnification structure image of a ring silicate (cordierite, $[Mg_{1.6}Fe_{0.3}Al_{0.1}]\ Al_4Si_5O_{18}$) is given in Figure 5. The

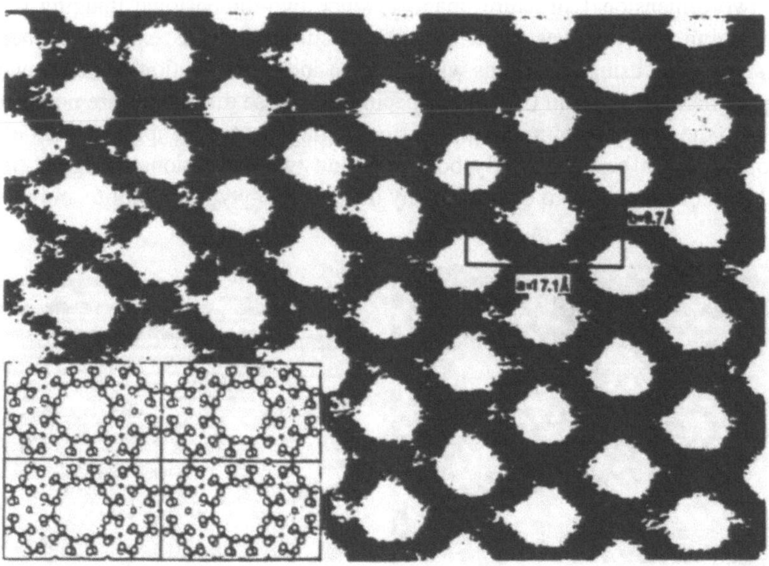

Figure 5. High-magnification structure image (two-dimensional) of cordierite. Dark areas corespond to region of high atomic density and white areas to tunnels with low atomic density. The unit cell is outlined. The insert is idealized cordierite structure [From Buseck and Iijima, 1974, *American Mineralogist,* 59:6–19, fig. 3b, copyright by the Mineralogical Society of America.]

inset shows the X-ray structure of several unit cells. The white areas in the HRTEM image represent large open channels. Dark areas of the image correspond to the positions of the high-atomic-density tetrahedrally coordinated Al and Si atoms cross linking $Al_2Si_4O_{18}$ rings. Structure imaging of muscovite did not give details as fine as those of cordierite, but the image formed by intersecting lattice fringes correctly exhibited crystal periodicities [Figure 6(a)]. One-dimensional lattice fringe imaging of the same muscovite showed 10-Å periodicities along the *c* axis corresponding to 1M polymorph [Figure 6(b)]. Yoshida (1976) examined the clay material which was the weathering product of granite (Ohya-ishi) and observed lattice images that are smaller than 10 Å, corresponding to d(002) and d(003) of mica crystal, and a crossed lattice image of 4.5-Å spacing corresponding to the (110) and (020) planes of montmorillonite surrounding a mica crystal. From this study he was able to deduce that the order in the stacking of layers is influenced by the distribution of cations in the interlayers but not by the ferrous–ferric transition in the octahedral sheet.

Iijima and Buseck (1978) studied the order and disorder in mica structures by one-dimensional (lattice fringe) and two-dimensional (structural fringe) imaging. They suggested that the positions of the tunnels, as seen in two-dimensional images, lying between the alkali ions, can describe well the disordered sequences in mica layers. Similar information on disordered sequences in layers (stacking disorder) revealing different polytypes was also observed by one-dimensional imaging. Therefore they suggested that one-dimensional lattice fringe imaging may be a more practical way of analyzing stacking disorder in layer silicates than the two-dimensional structure imaging, since two-dimensional imaging is time consuming and involves experimental difficulties such as exact alignment of crystal and focusing, problems with electron-induced radiation damage, and the requirement of very thin (<150 Å) specimens. These difficulties are not encountered to the same extent in one-dimensional imaging, and samples as thick as 500 Å or more can be imaged. By both one- and two-dimensional imaging, Iijima and Buseck were able to observe many different polytypes in biotite and musco-

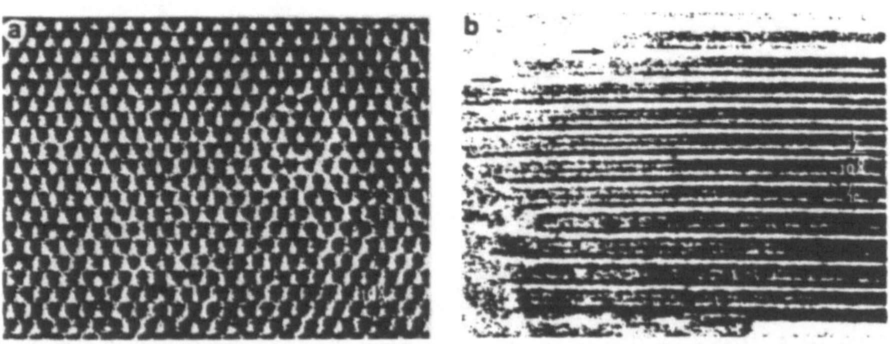

Figure 6. (a) High-magification structure image and (b) one-dimensional lattice fringe image of muscovite showing 10-Å periodicities along *c* axis corresponding to 1M polymorph [From Buseck and Iijima, 1974, *American Mineralogist,* 59:6–19, fig. 17, copyright by the Mineralogical Society of America.]

vite close to one another as coherent intergrowths with widths of a single unit cell [Figure 7(a) and (b)]. Mixed layering disorder between mica and chlorite was observed by Veblen (1983) [Figure 7(c)]. The stacking disorder of chlorite, revealing several different possible polytypes in chlorite, has also been studied using HRTEM (Spinnler et al., 1984). They indicated, however, that HRTEM should be combined with XRD analysis to obtain a complete picture of polytypes in chlorite.

Buseck and Veblen (1981) observed numerous varieties of columnar and planar defects in chain and sheet silicates, respectively. They clearly observed the interface between amphibole and talc, where amphibole was being replaced by talc (Figure 8). Iijima and Zhu (1982) also studied the interface between muscovite and biotite by HRTEM. They were able to observe the continuity in the interlayer and the octahedral sheet across the interface while 60% of the discontinuities were observed in the tetrahedral sheet. Other interesting features observed by Buseck and Veblen (1981) were the development of many types of

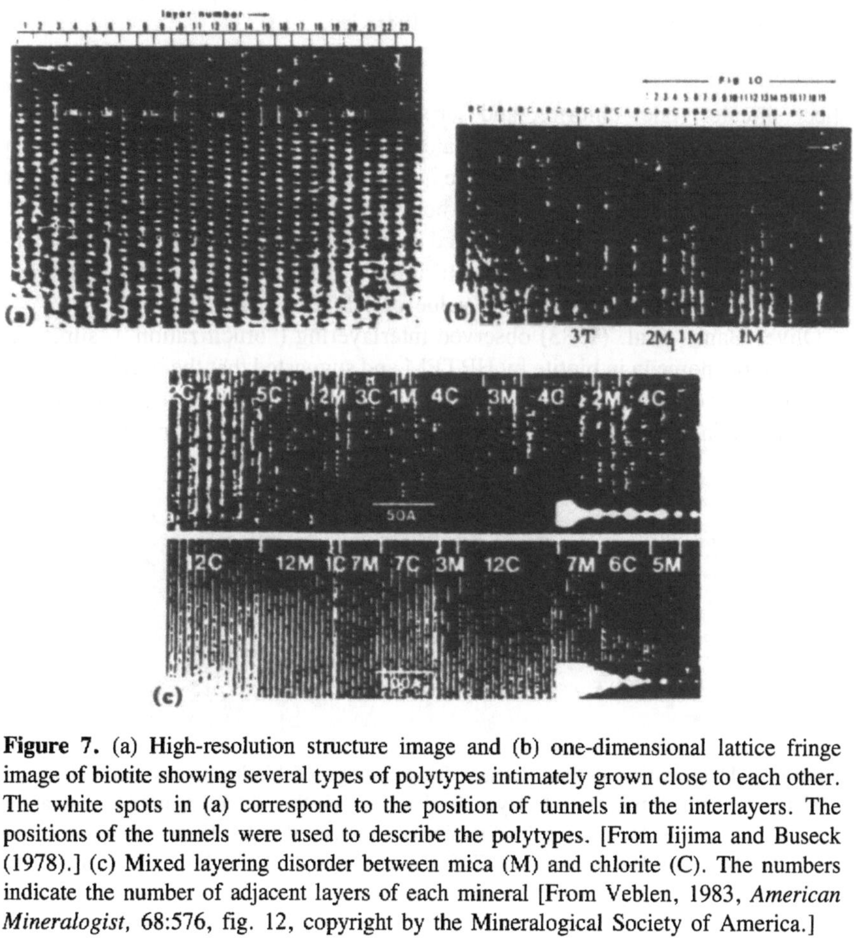

Figure 7. (a) High-resolution structure image and (b) one-dimensional lattice fringe image of biotite showing several types of polytypes intimately grown close to each other. The white spots in (a) correspond to the position of tunnels in the interlayers. The positions of the tunnels were used to describe the polytypes. [From Iijima and Buseck (1978).] (c) Mixed layering disorder between mica (M) and chlorite (C). The numbers indicate the number of adjacent layers of each mineral [From Veblen, 1983, *American Mineralogist,* 68:576, fig. 12, copyright by the Mineralogical Society of America.]

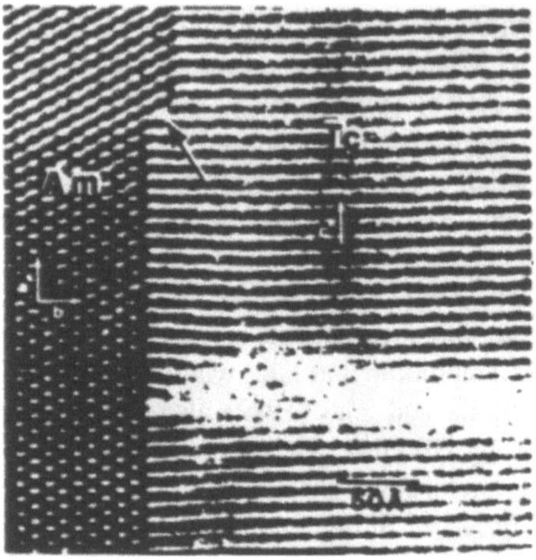

Figure 8. Amphibole (Am) being replaced by talc (Tc). The small ledge in the interface at the top of the figure is two amphibole chains wide (arrow) and migrates to give coherent growth of talc. [From Buseck and Veblen (1981).]

sheet silicates (talc, chlorite, and serpentine) from amphibole as a mixture. Sporadically, an entrapment of one talc layer between chlorite layers was also observed. Similar occlusion of chlorite in biotite was observed by Iijima and Zhu (1982), and occlusion of mica layers between kaolinite layers was observed by Lee et al. (1975), which they concluded as being a result of incomplete weathering of mica to kaolinite. Such an occlusion was not detected by XRD methods, and this is a clear example of the value of HRTEM over XRD.

Olives Banos et al. (1983) observed interlayering ("brucitization"), slip, and cleavage phenomena in biotite by HRTEM and suggested that the brucitization of the interlayer of biotite was facilitated by a planar defect, that is, partial basal slip or cleavage in the interlayer (Figure 9). According to them, the interlayer slip in

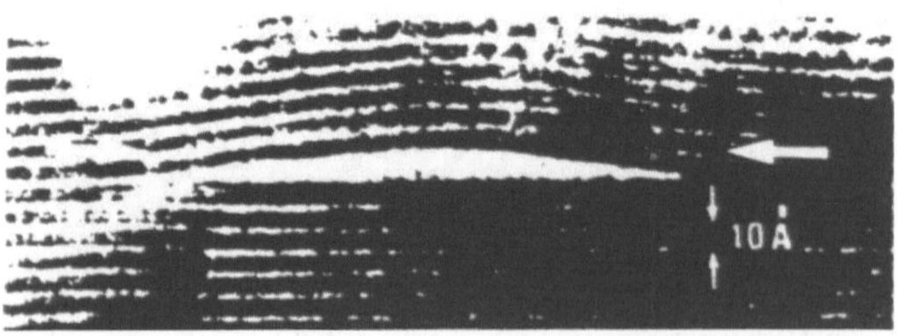

Figure 9. Deformation defect. Lattice fringe image of biotite shows microcleavage produced by basal slip in the interlayer. Such a defect is thought to promote brucitization of mica interlayer by facilitating the diffusion of cations and water. [From Olives Banos et al., 1983, *American Mineralogist*, 68:755, fig. 2, copyright by the Mineralogical Society of America.]

isolated areas of mica produces an expansion of the interlayer. This causes a reduction in cohesion energy between the layers, initiating the diffusion of "brucite-forming" materials in the interlayers. Amouric and Parron (1985) followed the structure and growth mechanism of a glauconite by HRTEM and concluded that the formation of this glauconite was an evolutionary process in which a gellike material transformed to 10-Å glauconite via 12.5-Å phase. They also observed the stacking defects in this glauconite, giving rise to different polytypes (1M, 1Md) including 2M1, which was reported for the first time in glauconite. Further details on the growth and deformation defects in phyllosilicates were reported by Amouric (1987).

Thus HRTEM provides detailed information on the structure and various types of defects on a unit-cell level of minerals. A direct observation of the defects in minerals leads to a better understanding of their transformation mechanisms and physical and chemical behavior, which is discussed in the subsequent sections. Even though HRTEM is an unrivaled method for studying intimate structures of crystals, Amouric et al. (1981) caution that the image formed by HRTEM may not be a true representation of, or a one-to-one correspondence to, the real structure in the crystal if the experimental conditions are not carefully controlled. They suggest that the calculations of images by computer will greatly aid in the interpretation of experimental results of localized defects in mica structures.

B. Mineral Transformation

In the natural environment, rocks and minerals are constantly altered from one phase to one or many other phases due to the forces of weathering. Alteration chiefly follows two reaction pathways: solid-state reaction or dissolution of the parent mineral and precipitation of a new phase or phases. Altered products may inherit most of the structural properties from the parent mineral, or they may exhibit entirely different structures leading to different physical and chemical properties. Therefore it is necessary to understand precisely the mechanism of mineral transformation to new or transitional phases. Although chemical, petrographic, and X-ray methods are useful in identifying reactants and products, they do not give a precise structural mechanism of reaction. HRTEM, in recent years, has proven to be invaluable in pinpointing the precise mechanisms of mineral alteration due to its ability to capture each stage of a reaction. Examples given in the following paragraphs demonstrate the usefulness of HRTEM in the study of mineral transformation.

Page and Wenk (1979) followed the hydrothermal alteration of plagioclase in a sulfide-bearing vein and confirmed that this feldspar first altered to smectite and then to 2M1 sericite through a series of highly disordered phyllosilicate intermediates. Eggleton and Buseck (1980) also studied the feldspar (microcline and plagioclase) weathering by HRTEM in humid, temperate climate and observed that the 10-Å layer silicate (illite and/or collapsed montmorillonite) was the only crystalline product of weathering in this environment. From this study they concluded that the dissolution of feldspar begins at the boundary of twinned and untwinned domains, leading to the production of circular holes where amorphous, ring-shaped structures crystallize to 10-Å layer silicate with irregular layer stacking (Figure 10).

Veblen and Ferry (1983), Olives Banos and Amouric (1984), and Olives Banos (1985) showed that the mechanism of transformation of biotite to chlorite occurs in two ways. Veblen and Ferry (1983) consider the chief mechanism of biotite⟶ chlorite to be a result of the stripping of tetrahedral sheets and K interlayers on either side of some of the 2:1 layers of biotite, leaving behind an octahedral sheet sandwiched between two 2:1 layers (mechanism 2) and giving rise to a chlorite-type structure [Figure 11(a)]. In the study of alteration of granitic biotite, Eggleton and Banfield (1985) also concluded that the main mechanism of biotite alteration to chlorite was mechanism 2 since the volume change in the alteration reactions could be accounted for by this mechanism. On the other hand, Olives Banos and Amouric (1984) and Olives Banos (1985) observed that the chloritization of biotite occurs by the replacement of K interlayers by brucitelike interlayers, giving rise to an ordered and disordered biotite–chlorite structure (mechanism 1). Such a brucitization is thought to be initiated by partial slip or cleavage in the interlayer level [Figures 9 and 11(b)]. A schematic representation of these two mechanisms is given in Figure 11(c).

In soils and sediments, micas (muscovite, biotite, and phlogopite) have long been thought to weather to smectite via vermiculite by replacement of interlayer potassium by hydrated cations, such as Mg^{2+} and Na^{+}, with a concomitant loss

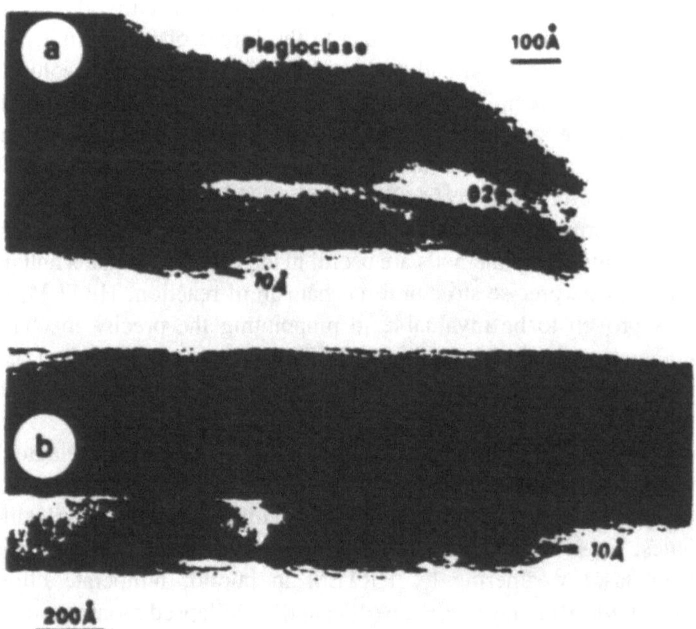

Figure 10. Feldspars being transformed to 10-Å layer silicate. (a) Plagioclase. (b) Microcline. [From Eggleton and Buseck (1980).]

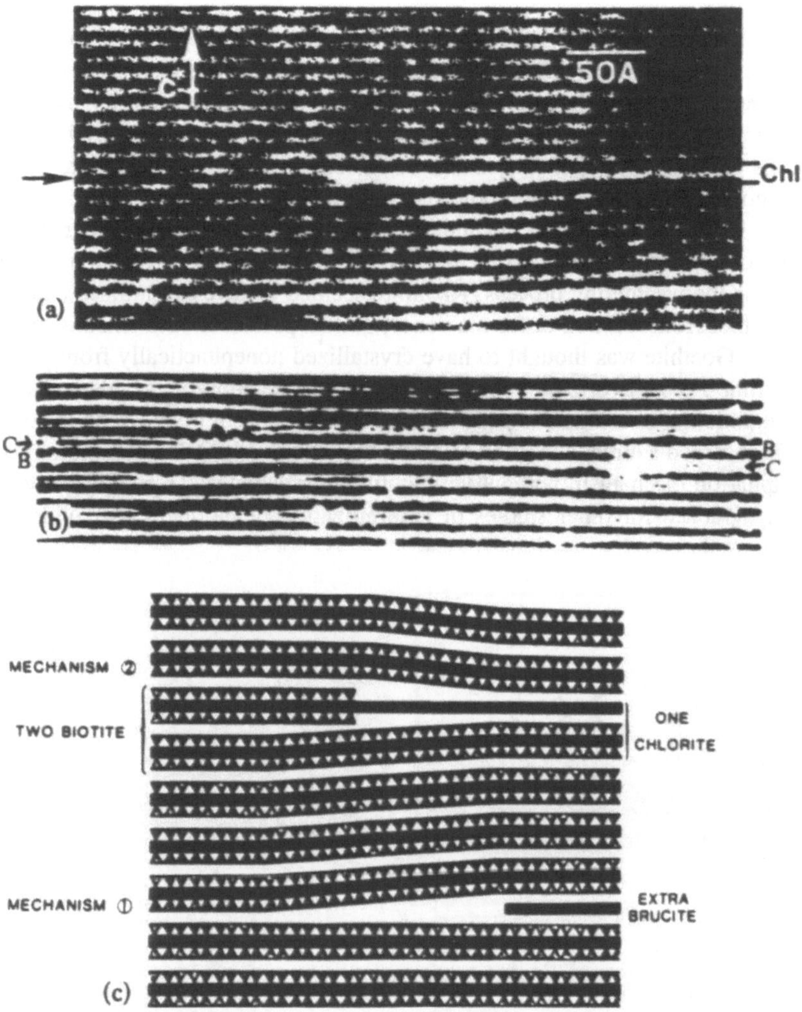

Figure 11. Transformation of biotite to chlorite. (a) Formation of one chlorite layer from two biotite layers by stripping of tetrahedral sheets and K interlayers (mechanism 2). [From Veblen and Ferry, 1983, *American Mineralogist,* 68:1162, fig. 3, copyright by the Mineralogical Society of America.] (b) Formation of one chlorite (C) layer by replacement of one K interlayer of biotite (B) by one brucitelike interlayer (mechanism 1). [From Olives Banos (1985).] (c) Schematic representation of mechanisms 1 and 2. [From Eggleton and Banfield, 1985, *American Mineralogist,* 70:907, fig. 8, copyright by the Mineralogical Society of America.]

of layer charge (Jackson et al., 1952; Norrish, 1973; Sridhar and Jackson, 1974). Not until recently, however, was direct evidence of these processes established. In the lattice images of soil dioctahedral vermiculites, Vali and Koster (1986) observed a core zone of ordered 10-Å layer spacing with increasingly disordered layers and increased layer spacing outward. These vermiculites could well be the product of muscovite weathering. In more recent studies, Banfield and Eggleton

(1988) tracked down by HRTEM the mechanisms of alteration of biotite to vermiculite, and subsequently to kaolinite and geothite, with increased weathering. They concluded that the biotite alters to vermiculite primarily by direct structural modifications, in which interlayer potassium is replaced by hydrated cations [Figure 12(a) and (b)] and where, less frequently, two biotite sheets undergo redistribution of components to form one vermiculite layer [Figure 12(c)]. TEM, however, did not indicate clearly whether the K^+ was replaced before or after the oxidation of Fe and whether the oxidation was responsible for initiating the weathering process. Kaolinite seemed to have crystallized epitactically either from solution or from a noncrystalline precursor onto existing biotite surface. Goethite was thought to have crystallized nonepitactically from poorly crystalline Fe oxyhydroxides formed from the segregation of Fe released from the degradation of biotite and vermiculite. Ahn and Peacor (1987a), on the other hand, observed a direct alteration of biotite to kaolinite through a dissolution and precipitation mechanism without vermiculite intermediates. This study provides direct evidence for other studies of kaolinitization of biotite, either under in-

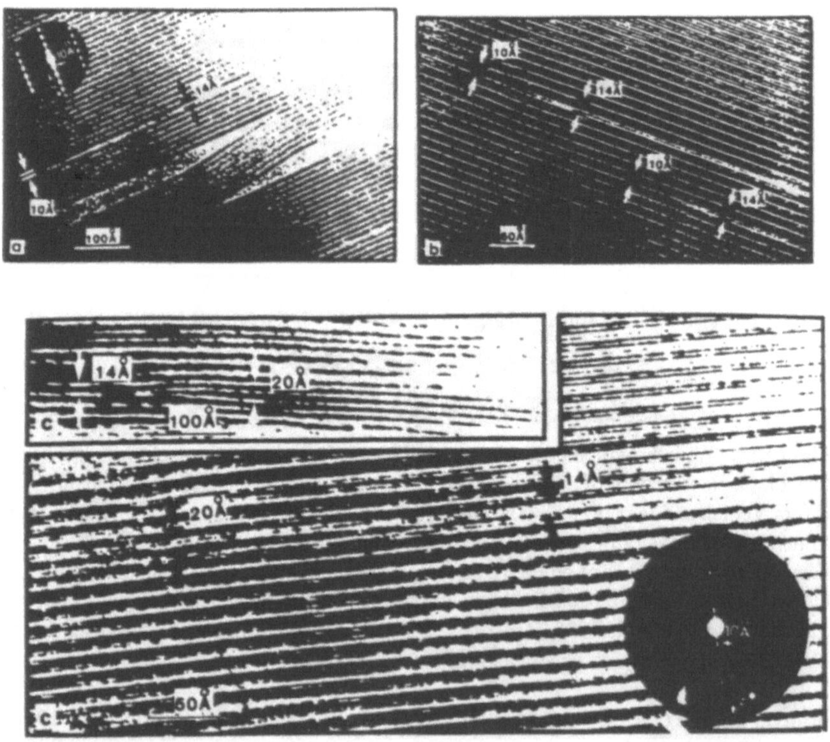

Figure 12. Transformation of biotite to vermiculite. (a), (b) formation of one 14-Å vermiculite layer from one 10-Å biotite layer by direct structural modification. (c) Formation of one vermiculite layer (14 Å) from two biotite layers (20 Å). [From Banfield and Eggleton 1988).]

tensive cropping (Rafols and Douglas, 1987) or under virgin conditions (Harris et al., 1985).

In sedimentary, metamorphic, or hydrothermal environments, smectite minerals are thought to have transformed to illite (illitization) through interstratified illite–smectite intermediates. Two possible mechanisms of smectite illitization have been considered. They involve either solid-state (direct transformation of layers with minor or major reorganization of elements in the layers) or dissolution and precipitation reactions, depending on the chemistry of the environments in which the transformation reactions are occurring. Both of these reactions, as approximated earlier by chemical and X-ray studies, have been supported by direct evidence produced by recent HRTEM studies. Bell (1986) observed a lateral termination of smectite layer to illite layer and therefore concluded that the lateral replacement of smectite by illite at edge dislocations is the principal growth mechanism. As noted by Bell and other workers, elements such as Al, Si, Fe, Ca, Mg, and Na were mobile along dislocations in order to facilitate the illitization. Ahn and Peacor (1986a) observed packets of defect-free illite within a matrix of smectite with imperfect and discontinuous layers. Therefore they postulated that the illitization proceeded as a result of the breakdown of major parts of octahedral and tetrahedral sheets of smectite facilitated by water. Real schematic representations of the transformation of smectite to illite in Gulf coast argillaceous sediments are given in Figure 13.

In another study of sediments of the Gulf coast, Ahn and Peacor (1987b) observed packets of kaolinite within the matrix of smectite or illite and typically interstratified with illite. From their HRTEM study they concluded that the kaolinite (in K-deficient environment) and illite simultaneously crystallized from the dissolution of smectite. Yau et al. (1988) observed discrete euhedral to subhedral crystals of illite, biotite, and chlorite in the argillaceous sediments of the Salton Sea, California, and concluded that these phyllosilicates formed through nucleation and crystal growth (from dissolved smectite) without intermediate phases rather than by direct replacement of smectite.

HRTEM, thus, captures each stage of the mineral transformation and aids in understanding their mechanisms. Amouric et al. (1981), however, cautioned that the images formed by HRTEM may sometimes be misleading if the experimental conditions are not carefully controlled. Ahn and Peacor (1986b) demonstrated both 20-Å and 10-Å periodicities for a rectorite sample at two different focus conditions.

C. Fundamental Particle and Interstratified Minerals

Interstratified clay minerals are widespread in both soils and sediments. A mineral is considered interstratified when two or more types of silicate layers are arranged in a regular, partially ordered, or random fashion, which makes up a mosaic of discrete domains within a crystallite. Most of the interstratified clays involve mica–smectite (rectorite), illite–smectite, mica–vermiculite, chlorite–smectite (corrensite), and kaolinite–smectite. Interstratifications in minerals are

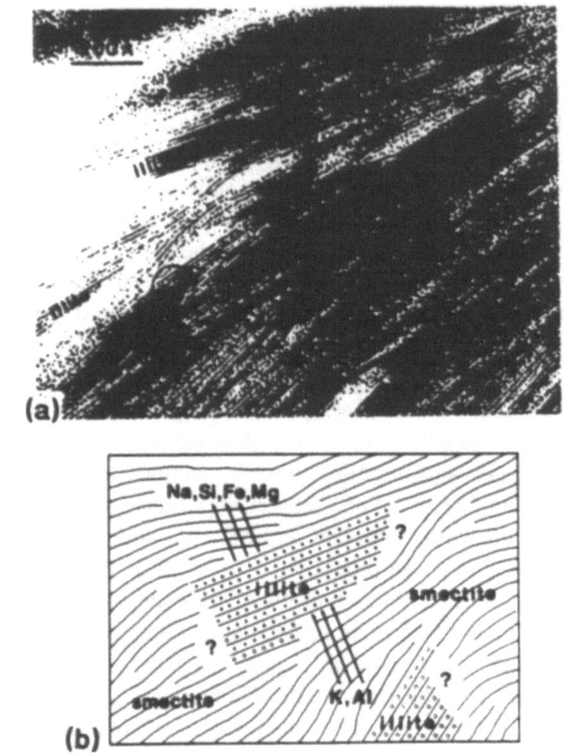

Figure 13. Transformation of smectite to illite. (a) Formation of thin packets of straight, relatively defect-free illite layers within highly imperfect and wavy matrix of smectite. (b) Model of smectite illitization. [From Ahn and Peacor (1986a).]

commonly studied by XRD. However, recent studies of interstratification in clays employing TEM (Nadeau et al., 1984a, 1984b, 1984c; Nadeau, 1985; Wilson and Nadeau, 1985; Wilson, 1987) have questioned the existing concept of interstratification as deduced from XRD. According to these authors, the transformation of minerals in weathering environment (soils) that involves relatively large crystals of mica and chlorite to interstratified vermiculite and then to smectite is a true interstratification. Interstratified minerals formed during diagenesis that involve extremely fine-grained particles, however, are not a true interstratification. In this case, the interfaces between thin particles of illite (elementary particle) can adsorb water and organic molecules (ethylene glycol or glycerol), and these thin particles can exhibit interparticle diffraction. Therefore XRD perceives them as smectite. TEM studies showed that 10-Å-thick particles correspond to a single 2:1 layer of smectite, 20-Å-thick particles to elementary illite (two 2:1 layers sandwiching a plane of potassium ions), and 24-Å-thick particles to elementary chlorite (two 2:1 silicate layers plus a single brucite layer) [Figure 14(a)].

From these observations Nadeau et al. (1984) postulated that regularly in-

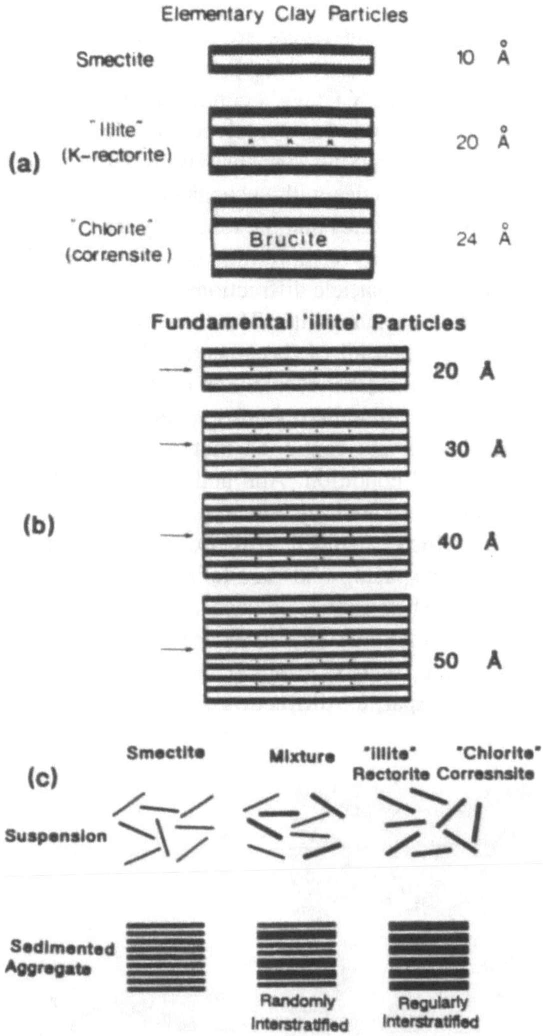

Figure 14. Schematic representations. (a) Elemetary clay particles. (b) Fundamental illite particles. (c) Elemetary clay particles in suspension and after sedimentation. [From Nadeau et al. (1984c).]

terstratified illite–smectite (K rectorite) and chlorite–smectite (corrensite) are composed of particles of 20-Å-thick illite and 24-Å chlorite, respectively; other interstratified minerals, such as illite–smectite of the IIS and IIIS types are composed of particles (illite) 20 to 50 Å thick [Figure 14(b)]. However, these minerals appear interstratified with smectite because the interfaces of these minerals are capable of adsorbing water or organic molecules and are capable of interparticle diffraction. Interparticle diffraction is obtained from the sedimented aggregate of elementary particles dried on the flat glass slide since the effective number of coherently diffracting layers are increased due to the parallelism between the nearby particles [Figure 14(c)]. These elementary particles otherwise would not be able to satisfy Bragg's condition to produce any diffraction. Randomly interstratified minerals (as perceived by XRD), on the other hand, were synthesized simply by mixing suspensions of 20-Å illite (rectorite) and 24-Å chlorite (corrensite) particles with suspensions of 10-Å smectite particles [Figure 14(c)], which demonstrates that the randomly interstratified minerals are simply physical mixtures of two or more minerals. Illite particles when thicker than 50 Å are capable of intraparticle diffraction producing 10-Å basal spacings, and the clay is identified as pure illite. This is a common observation in the diagenetic environment, when all of the smectite converts to illite and grows after dissolution and crystallization. Mackinnon (1987) argued that the existence of elementary particles of illite in regularly interstratified illite–smectite (K rectorite) is not conclusive and cautioned that such observation could be an artifact of the experimental condition. Ahn and Peacor (1986b) indeed showed the existence of randomly oriented fundamental 20-Å-thick particles of rectorite [Figure 15(a) and (b)] which expanded with dodecylammonium ion, giving rise to 12- to 15-Å interlayer spacing and parallelism between a few fundamental particles [Figure 15(c)]. They added, however, that these particles might have been produced by preferential breakage of rectorite along smectitelike interlayers during sample preparation, that is, grinding with mortar and pestle.

Thus the concept of interparticle diffraction of fundamental particles to explain

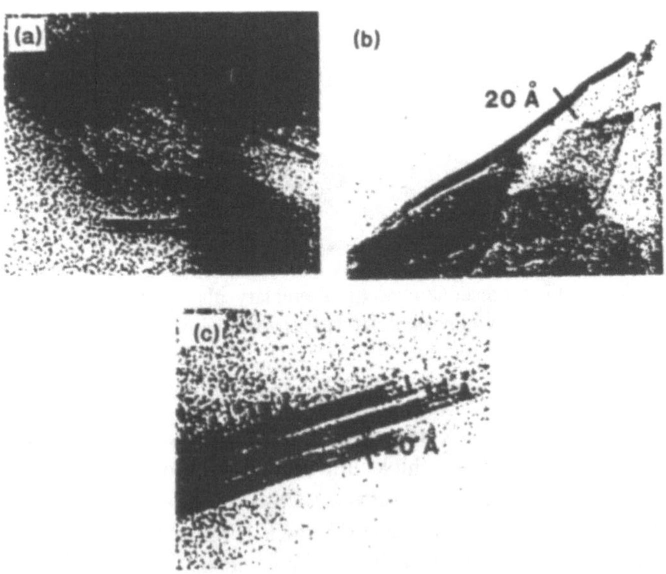

Figure 15. Lattice fringe images of rectorite showing existence of fundamental particles. (a) Short 20-Å particles. (b) Long 20-Å particle. (c) After dodecylammonium ion intercalation. [From Ahn and Peacor (1986b).]

the interstratification in minerals was made possible by direct observation of the physical dimensions of fundamental particles. The new concept of fundamental particle for interstratified minerals is thought to have several implications. This model suggests the neoformation or diagenetic conversion of smectite to illite rather than a simple transformation reaction. For thermodynamic purposes, the regularly interstratified illite–smectite or chlorite–smectite could be considered a single-phase illite or chlorite, and randomly interstratified minerals could be considered mixtures of two phases. Although interstratified minerals in soils can be regarded as single phases (Wilson and Nadeau, 1985), the validity of the new concept of fundamental particles and interstratification remains to be shown in numerous situations.

D. Physical and Chemical Properties

HRTEM is useful for the direct observation of the structural makeup and various defects in mineral samples and for the identification of each phase of the mineral alteration reactions, which helps to predict and understand their physical and chemical behavior. HRTEM has also aided greatly in unraveling some of the thermodynamic and kinetic problems of minerals which previously were not clearly understood by conventional methods.

Clay texture is a very important parameter that controls both the structure and the hydration behavior of the soil system. In the studies of smectite by HRTEM complemented by small-angle X-ray scattering (SAXS), Ben Rhaiem et al. (1987) showed that the microstructure (layer spacings and layer organization) of the clay is strongly affected by the types of cations present and the history of applied pressure. Since the applied pressure tends to increase the number of layers per particle of smectite, drying and wetting involve changes in both particle size and the internal structure of the particle. For a Ca smectite prepared under low suction pressure (high water content), the pore was formed from the organization of quasicrystals made up of several layers (≈50) arranged face to face with great lateral extension, where internal discontinuities as large as 50 Å were observed [Figure 16(a)]. At high water content, the pore water accounted for about 92% of the sample water since the layer would not expand more than 18 Å. A pressure greater than 50 bar had to be applied to decrease the basal spacing of Ca smectite from 18 to 15 Å. With successive drying, the system lost water and the pore walls became thicker. For Na smectite under low suction pressure, the particle (tactoids) was made up of fewer layers (4 to 10) [Figure 16(b)] than for Ca smectite and the interlayer distance varied considerably depending on the suction pressure. The internal arrangement of tactoids observed by HRTEM and verified by SAXS showed that for $p > 25$ bar, the substacks (numbers of layers $\leqq 4$) had interlayer spacings of 35 Å, and for $p < 10$ bar, the interlayer spacings were $\geqq 35$ Å (with some interlayers as high as 100 Å), indicating the existence of diffuse double layers. At this stage the interlayer water may account for most of the soil water. Schematic representations of Ca and Na montmorillonite

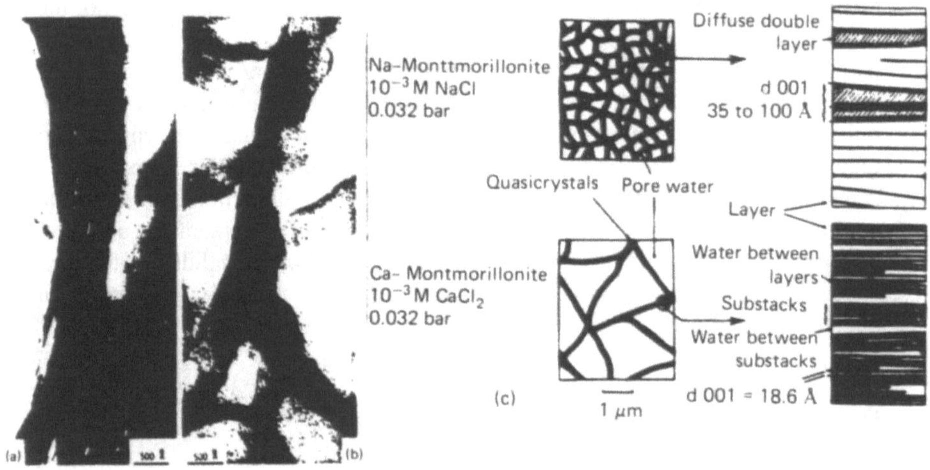

Figure 16. Lattice fringe images showing layer organizations in clay–water system. (a) Ca smectite. (b) Na smectite. (c) Schematic representation of microstructures Ca and Na montmorillonite with dilute salt solutions (10^{-3} *M*). [From Ben Rhaiem et al. (1987).]

as observed by HRTEM and SAXS are given in Figure 16(c). Robert (1987) indicated that the phenomena described with regard to the changes in the layer organization of smectite clays with applied pressure or water content may be important in the diffusion and retention of cations in the soil system. From the evidence obtained from HRTEM and scanning electron microscopy (SEM), Tessier and Pedro (1988) showed that the type of clays and the rate of rewetting are also important in the stability of the soil structure. Both HRTEM and SEM contributed to determine the rigidity of clay particles, pore size, and pore continuity, all of which are important for the stability of the clay structure.

Pickles and Schlup (1985) determined the particle association in the suspension of two synthetic smectites. They observed the particles as isolated platelets or aggregates of platelets three to five unit cells large. Determination of the physical dimensions of clay particles has great significance in clay colloid chemistry for a variety of industrial applications (Pickles and Schulp, 1985; Nadeau, 1985).

HRTEM permits the direct observation of heterogeneity in clays. The heterogeneities may be from particle to particle or within a particle. Some of them are described in Sections IV-A and B. For example, feldspars may weather to smectite and illite or sericite (Eggleton and Buseck, 1980; Page and Wenk, 1979) and biotite to chlorite (Veblen and Ferry, 1983) or to vermiculite, kaolinite, and goethite (Banfield and Eggleton, 1988) and thus form a heterogeneous assemblage of minerals. The weathered assemblages will exhibit a different chemistry from those that are not weathered. The weathering of both feldspar and biotite has important consequences in the potassium dynamics in a soil system since these minerals release potassium as they weather. The observation of heterogeneity in minerals by HRTEM within a particle, such as the presence of both 10-Å (mica-type) and 14-Å (vermiculite, smectite, or chlorite) layers or the occlusion of some layers (mica) in a thick matrix of other (kaolinite) layers (Lee et al., 1975), has provided insight in understanding their chemical behavior (such as ion exchange and release and fixation of ions). Robert (1987) pointed out that the presence of both 10- and 14-Å layers in a particle may lead to difficulties in predicting soil fertility on the basis of total mineralogical composition, since potassium could be released from one type of layer (10 Å) and fixed by the other (14 Å).

The heterogeneity of charge within a layer of clay has also been observed with the help of HRTEM (Vali and Koster, 1986; Yoshida and Suito, 1972). Such local variations within a layer may result in the fixation of cations that is stronger in one part than in the other.

VI. Conclusions

HRTEM has emerged as a powerful tool for the study of minerals. It can resolve a unit smaller than 3 Å and therfore is suitable for lattice imaging of minerals, especially layer silicates. It aids in the direct observation of structural features, including various defects, which is not possible by XRD, and thus allows one to follow each stage of mineral transformation. It also gives direct information in minute detail on the physical dimensions and layer organization of minerals, and permits predicting their physical and chemical behavior under various environmental conditions (such as history of applied stress and solution chemistry).

One should, however, be cautious in the interpretation of high-resolution images, because the images formed may not always be a true representation of the specimen due to their sensitivity to specimen thickness and the degree of defocus. Two different images may be observed at different focus conditions. It is therefore necessary to calculate a series of images at different specimen thicknesses and amounts of defocus, and to compare them with a series of observed images, especially while interpreting two-dimensional images. Other factors that affect the image contrast are high vacuum, irradiation (heat) damage, and various aberrations present in the lenses.

The knowledge gathered from the studies of individual clay minerals by HRTEM has yet to be fully applied to the processes occurring in soil systems. HRTEM may be a viable approach to study quantitatively the role of soil clay minerals in controlling K dynamics, to understand the hydration and structure of soils, and to study the transformation of minerals under various cropping systems or with other human intervention in soils.

References

Agar, A. W. 1978. Operation of the electron microscope. pp. 166–190. In: A. W. Agar, R. H. Alderson, and d. Chescoe. (eds.), *Principles and practice of electron microscope*. North-Holland, Amsterdam.

Amouric, M. 1987. Growth and deformation defects in phyllosilicates as seen by HRTEM. *Acta Cryst.* B43:57–63.

Amouric, M., and C. Parron. 1985. Structure and growth mechanism of glauconite as seen by high-resolution transmission electron microscopy. *Clays Clay Mineral.* 33:473–482.

Amouric, M., G. Mercuriot, and A. Baronnet. 1981. On computed and observed HRTEM images of perfect mica polytypes. *Bull. Mineral.* 104:298–313.

Ahn, J. H., and D. R. Peacor. 1986a. Transmission and anlytical electron microscopy of the smectite-to-illite transition. *Clays Clay Mineral.* 34:165–179.

Ahn, J. H., and D. R. Peacor. 1986b. Transmission electron microscope data for rectorite:Implications for the origin and structure of fundamental particles". *Clays Clay Mineral.* 34:180–186.

Ahn, J. H., and D. R. Peacor. 1987a. Kaolinitization of biotite: TEM data and implications for an alteration mechanism. *Amer. Mineral.* 72:353–356.

Ahn, J. H., and D. R. Peacor. 1987b. Transmission electron microscopy study of the diagenesis of kaolinite in Gulf Coast argillaceous sediments. pp. 151–157. In: L. G. Schultz, H. van Olphen, and F. A. Mumpton. (eds.), *Proc. Int. Clay Conf., Denver, 1985*. Clay Minerals Soc., Bloomington, IN.

Allpress, J. G., and J. V. Sanders. 1973. The direct observation of the structure of real crystals by lattice imaging. *J. Appl. Cryst.* 6:165–190.

Banfield, J. F., and R. A. Eggleton. 1988. Transmission electron microscope study of biotite weathering. *Clays Clay Mineral.* 36:47–60.

Bell, T. E. 1986. Microstructure in mixed-layer illite/smectite and its relationship to the reaction of smectite to illite. *Clays Clay Mineral.* 34:146–154.

Ben Rhaiem, H., C. H. Pons, and D. Tessier. 1987. Factors affecting the microstructure of smectites:role of cation and history of applied stresses. pp. 292–297. In: L. G. Schultz, H. van Olphen, and F. A. Mumpton. (eds.), *Proc. Inter. Clay Conf., Denver, 1985*. Clay Minerals Soc., Bloomington, IN.

Brown, J. L., and M. L. Jackson. 1973. Chlorite examination by ultramicrotomy and high resolution electron microscopy. *Clays Clay Mineral.* 21:1–7.

Brown, J. L., and C. I. Rich. 1968. High-resolution electron microscopy of muscovite. *Science* 161:1135–1137.

Buseck, P. R., and J. M. Cowley. 1983. Modulated and intergrowth structures in minerals and electron microscope methods for their study. *Amer. Mineral.* 68:18–40.

Buseck, P. R., and S. Iijima. 1974. High resolution electron microscopy of silicates. *Amer. Mineral.* 59:1–21.

Buseck, P. R., and D. R. Veblen. 1981. Defects in minerals as observed with high-resolution transmission electron microscopy. *Bull. Mineral.* 104:249–260.

Comer, J. J. 1971. Specimen preparation. pp. 79–108. In: J. A. Gard (ed), *The electron-optical investigation of clays*. Mineral. Soc., London.

Cowley, J. M., and S. Iijima. 1976. The direct imaging of crystal structures. pp. 123–136. In: H.-R. Wenk. (ed.), *Electron microscopy in mineralogy*. Springer-Verlag, Berlin.

Edington, J. W. 1975. Interpretation of transmission electron micrographs, Monograph 3. In: *Practical electron microscopy in materials science*. Macmillan, London.

Eggleton, R. A. 1988. The application of micro-beam methods to iron minerals in soils. pp. 165–201. In J. W. Stucki, B. A. Goodman, and U. Schwertmann. (eds.), *Iron soils and clay minerals*. Ridel, Boston.

Eggleton, R. A., and J. F. Banfield. 1985. The alteration of granitic biotite to chlorite. *Amer. Mineral.* 70:902–910.

Eggleton, R. A., and P. R. Buseck, 1980. High resolution electron microscopy of feldspar weathering. *Clays Clay Mineral.* 28:173–178.

Harris, W. G., L. W. Zelazny, and F. D. Bloss. 1985. Biotite kaolinitization in Virginia Piedmont soils: II. Zonation in single grains. *Soil Sci. Soc. Amer. J.* 49:1297–1302.

Hirsch, P. B. A., A. Howie, R. B. Nicholson, D. W. Pashley, and M. J. Whelan. 1965. *Electron microscopy of thin crystals*. Butterworths, London.

Iijima, S., and J. G. Allpress. 1974. Structural studies by high-resolution electron microscopy: Tetragonal tungsten bronze-type structures in the system Nb_2O_5–WO_3. *Acta Cryst.* A30:22–29.

Iijima, S., and P. R. Buseck. 1978. Experimental study of disordered mica structures by high resolution electron microscopy. *Acta Cryst.* A34:709–719.

Iijima, S., and J. Zhu. 1982. Electron microscopy of a muscovite-biotite interface. *Amer. Mineral.* 67:1195–1205.

Jackson, M. L., Y. Hseung, R. B. Corey, E. J. Evans, and R. C. Vanden Heuvel. 1952. Weathering sequence of clay-size minerals in soils and sediments: II. Chemical weathering of layer silicates. *Soil Sci. Soc. Amer. Proc.* 16:3–6.

Lee, S. Y., M. L. Jackson, and J. L. Brown. 1975. Micaceous occlusions in kaolinite observed by ultramicrotomy and high resolution electron microscopy. *Clays Clay Mineral.* 23:125–129.

Mackinnon, I. D. R. 1987. The fundamental nature of illite/smectite mixed-layer clay particle: A comment on papers by Nadeau and coworkers. *Clays Clay Mineral.* 35:74–76.

Malla, P. B., M. Robert, L. A. Douglas, and d. Tessier. 1987. Charge heterogeneity and microstructure of 2:1 layer silicates. *Agron. Abstracts* p. 274.

McKee, T. R., and J. L. Brown. 1977. Preparation of specimens for electron microscopic examination. pp. 809–846. In: J. B. Dixon, and S. B. Weed. (eds.), *Minerals in soil environments*. Soil Sci. Soc. Amer., Madison, WI.

Nadeau, P. H. 1985. The physical dimensions of fundamental clay particles. *Clay Mineral.* 20:499–514.

Nadeau, P. H., and J. M. Tait. 1987. In M. J. Wilson, (ed.), *Handbook of determinative methods in clay mineralogy*. Blackie and Son, London.

Nadeau, P. H., M. J. Wilson, W. J. McHardy, and J. M. Tait. 1984a. Interstratified clays as fundamental particles. *Science* 225:923–925.

Nadeau, P. H., J. M. Tait, W. J. McHardy, and M. J. Wilson 1984b. Interstratified XRD characteristics of physical mixtures of elementary clay particles. *Clay Mineral.* 19:67–76.

Nadeau, P. H., M. J. Wilson, W. J. McHardy, and J. M. Tait. 1984c. Interparticle diffraction: A new concept for interstratified clays. *Clay Mineral.* 19:757–769.

Norrish, K. 1973. Factors in the weathering of micato vermiculite. pp. 417–432. In: J. M. Serratosa (ed.), *Proc. Int. Clay Conf., Madrid, 1972*. Div. Ciencias CSIC, Madrid.

Olives Banos, J. 1985. Biotites and chlorites as interlayered biotie-chlorite crystals. *Bull. Mineral.* 108:635–641.

Olives Banos, J., and M. Amouric. 1984. Biotite chloritization by interlayer brucitization as seen by HRTEM. *Amer. Mineral.* 69:869–871.

Olives Banos, J., M. Amouric, C. D. Fouquet, and A. Baronnet. 1983. Interlayering and interlayer slip in biotite as seen by HRTEM. *Amer. Mineral.* 68:754–758.
Page, R., and H. R. Wenk. 1979. Phyllosilicate alteration of plagioclase studied by transmission electron microscopy. *Geology* 7:393–397.
Paulus, M., A. Dubon, and J. Etienne. 1975. Application of ion-thinning to the study of the structure of argillaceous rocks by transmission electron microscopy. *Clay Mineral.* 10:417–426.
Pickles, D. G., and J. R. Schlup. 1985. Particle association in smectite soils by transmission electron microscopy. *Clays Clay Mineral.* 33:362–366.
Rafols, N. M., and L. A. Douglas. 1987. Transformation of glauconite to kaolinite under intensive cropping. p. 112. In *Abstracts 24th Ann. Meet. Clay Minerals Soc.*, Socorro, NM.
Robert, M. 1987. Some general aspects of K dynamics and new trends in soil mineralogy. In: *Trans. 13th Cong. Int. Soc. Soil Sci.*, Hamburg, pp. 1122–1132.
Smart, P., and N. K. Tovey. 1982. *Electron microscopy of soils and sediments: Techniques*. Clarendon, Oxford, 400 pp.
Spinnler, G. E., P. G. Self, S. Iijima, P. R. Buseck. 1984. Stacking disorder in clinochlore chlorite. *Amer. Mineral.* 69:252–263.
Spurr, A. R. 1969. A low-viscosity epoxy resin embedding medium for electron microscopy. *J. Ultrastruc. Res.* 26:31–43.
Sridhar, K., and M. L. Jackson. 1974. Layer charge decrease by tetrahedral cation removal and silicon incorporation during natural weathering of phlogopite to saponite. *Soil Sci. Soc. Amer. Proc.* 38:847–850.
Tessier, D., 1984. Etude expérimentale de l'organisation des matériaux argileux. D. es Sci. thesis, INRA, Paris, 361 pp.
Tessier, D., and G. Pedro. 1988. Influence of clay mineralogy and rewetting rate on clay structure. In: L. A. Douglas. (ed.), *Proc. Int. Working Meet. on Soil Micromorphology, San Antonio, TX, June 10–15, 1988*. Soil Sci. Soc. Amer., Madison, WI.
Vali, H., and H. M. Koster. 1986. Expanding behavior, structural disorder, regular and random irregular interstratification of 2:1 layer-silicates studied by high-resolution images of transmission electron microscopy. *Clay Mineral.* 21:827–859.
Veblen, D. R. 1983. Microstructure and mixed layering in intergrown wonesite, chlorite, talc, biotite, and kaolinite. *Amer. Mineral.* 68:566–580.
Veblen, D. R. 1985. Direct TEM imaging of complex structures and defects in silicates. *Ann. Rev. Planet. Sci.* 13:119–146.
Veblen, D. R., and J. M. Ferry. 1983. A TEM study of the biotite-chlorite reaction and comparison with petrologic observations. *Amer. Mineral.* 68:1160–1168.
Wenk, H. R. (ed.) 1976. *Electron microscopy in mineralogy*. Springler, Berlin.
Wilson, M. J. 1987. Soil smectites and related interstratified minerals: Recent developments. pp. 167–173. In: L. G. Schultz, H. van Olphen, and F. A. Mumpton (eds.), *Proc. Int. Clay Conf., Denver, 1985*. Clay Minerals Soc., Bloomington, IN.
Wilson, M. J., and P. H. Nadeau. 1985. Interstratified clay minerals and weathering processes. In: J. I. Drever (ed.), *The chemistry of weathering*. Reidel, Boston.
Yau, Y. C., D. R. Peacor, R. E. Beane, E. J. Essence, and S. D. McDowell. 1988. Microstructures, formation mechanisms, and depth-zoning of phyllosilicates in geothermally altered shales, Salton Sea, California. *Clays Clay Mineral.* 36:1–10.
Yoshida, T. 1976. Study of microstructure of mica and montmorillonite by high resolution electron microscopy. *Clay Sci.* 5:1–7.
Yoshida, T., and E. Suito. 1972. Interstratified layer structure of the organo-montmorillonites as releaved by electron microscopy. *J. Appl. Cryst.* 5:119–124.

Bare-Soil Evaporation in Relation to Tillage

S. K. Jalota and S. S. Prihar*

I. Introduction

Direct evaporation from soil constitutes a major pathway of water loss, which is considered wasteful since it does not contribute to crop production. It is estimated that from 50 to 70% of annual precipitation is returned to the atmosphere as direct evaporation without any benefit to crop production (Hide, 1954; Peters and Russel, 1959; Harrold et al., 1959; Peters, 1960). In fact, even in growing crops direct evaporation from (exposed) soil surface E_s accounts for a substantial part of evapotranspiration (ET). Tanner et al. (1960) estimated that in widely spaced row crops such as corn, if the soil is wetted frequently by rain or irrigation, E_s may be as high as 50% of ET, even when the canopy is fully developed. Obviously there is a need to reduce this loss since it will increase the

*Department of Soils, Punjab Agricultural University, Ludhiana 141004, India.

storage of plant-available water in the root zone in the bare soil and cause a greater fraction of crop ET to be used by transpiration. It is well known that under a given climate dry-matter production (de Wit, 1958; Arkley, 1963) and even grain production (Hanks et al., 1969a) is a linear function of transpiration by the crop. Upward movement of water due to evaporation not only reduces soil water storage but also deposits salts within the root zone.

Various mulches such as plastic films, paper mulch, gravel, and straw mulch, which are applied extraneously at the soil surface, are recommended for reducing evaporation. However, these mulches are not cost effective except for very high value crops. Soil mulch, created by shallow tillage after wetting and often followed by a smoothing operation, is a rather common practice aimed at the conservation of soil water, especially in the seed zone. Although the role of tillage in reducing water loss by destroying weeds is well recognized, its role in retarding evaporation under weed-free situations has been a matter of controversy for several decades. More recent investigations have shown that the time trend and the magnitude of evaporation reduction by shallow tillage are determined by relative drying of tilled vis-à-vis untilled soil, which depends on soil and climatic factors as well as time (after wetting) and type of tillage. This chapter discusses the dynamics of evaporation from bare and tilled soils, evaporation reduction by tillage, and profile distribution of water as affected by soil, climate, and tillage variables.

II. Process of Evaporation

Evaporation from soil or, for that matter, any other body is a two-step phenomenon, namely, conversion of liquid water to vapor and its escape into the atmosphere. Admittedly, there are three requisites for evaporation to proceed: (1) the source of energy to meet the latent heat of vaporization of water, (2) the existence of a vapor pressure gradient between the evaporating surface and the surrounding atmosphere, and (3) the supply of water to the evaporating surface (Hillel, 1971). The first two conditions, supply of energy and removal of vapor, are external to the evaporating body (soil) and are influenced by meteorological factors such as air temperature, wind velocity, and radiation, which together determine the evaporative demand, or potential evaporation, or atmospheric evaporativity. Hillel (1971) defined the atmospheric evaporativity E_0 as the maximim flux at which the atmosphere can vaporize water from a free water surface.

Evaporation from soil in the absence of a shallow water table is a transient process. Immediately following wetting by rain or irrigation the soil has a high water content and can conduct water at high rates. The water content of the surface layer and the water-content-dependent transmission coefficient decrease with time, but the potential gradient increases. Since the water flux is a product of the hydraulic conductivity of the soil and the hydraulic head gradient, which also changes as a function of time, it is natural for evaporation rates to change with time. Researchers (Stanhill, 1955; Lemon, 1956) have divided the process of evaporation from soil into three stages. In the first stage, the *constant-rate stage*, the evaporation rate is governed by external conditions and equals the atmospheric evaporativity. E_0 (Penman, 1941; Stanhill, 1955; Bond and Willis, 1970). But after a certain threshold period water loss from soil lags behind E_0 and decreases with time. This second stage, in which the evaporation rate decreases with time, is the *falling-rate stage*. Gradually a dry layer develops at the surface, which acts as barrier to vapor diffusion, and the evaporation enters the third stage, the *reduced-evaporation-rate stage*. Since the plane of evaporation shifts to a level below the surface, water transfer occurs only in the vapor state by molecular diffusion and the flow is governed by the adsorptive forces acting over molecular distances at solid–liquid interfaces.

Based on elaborate experiments Jackson et al. (1973) showed that under field conditions the concept of three stages of drying does not hold strictly. With diurnal fluctuations in the evaporative demand, during nighttime, when the temperature is low, surface layers absorb moisture because both water-content and temperature gradients are in a downward direction and cause water to flow to the surface (Rose, 1968). With the sun rise the evaporation rate equals E_0 for some time, and with a rise in temperature, the temperature gradient reverses and causes water vapor to move downward, with the result that the moisture content of the surface layer decreases and evaporation starts to occur as in the falling-rate stage. Minhas (1985) also could not discern three distinct stages under field conditions.

III. Factors Affecting Bare-Soil Evaporation

Important factors affecting the course of evaporation from bare soil are evaporativity, soil characteristics, and initial soil-water-content distribution in the profile.

A. Effect of Evaporativity

The effect of evaporativity E_0 on evaporation has been investigated by a number of workers. Penman (1941), Wiegand and Taylor (1960), Willis (1962), Gardner and Hillel (1962), Benoit and Kirkham (1963), Fritton et al. (1967), and Acharya et al. (1979) included more than one level of E_0 in their controlled experiments. Two divergent views emanated from these studies. Gardner and Hillel (1962) showed theoretically as well as experimentally that the cumulative evaporation (CE) under high E_0 remains always higher than that under lower E_0. The other point of view was that the initial higher evaporation rate under high E_0 may in the long run reduce total water loss to the atmosphere due to rapid formation of the dry layer, which reduces later loss of water (Covey and Bloodworth, 1966; Penman, 1941; Soviet work cited in Lemon, 1956). Strictly speaking the CE from soil is not only dependent on the magnitude of E_0 but also on the evaporative conditions and water retention and transmission properties of the soil.

The first concept was based on the solution of the general flow equation, Equation (1), given by Crank (1956) for the unidirectional movement of unsaturated flow, neglecting gravity,

$$\frac{\partial \theta}{\partial t} = \frac{\partial}{\partial z}\left[D(\theta)\ \frac{\partial \theta}{\partial z}\right] \tag{1}$$

Gardner (1959) solved Equation (1) for isothermal drying of semi-infinite homogeneous soil columns and obtained Equation (2),

$$\mathrm{CE} = 2(\theta_i - \theta_0)\sqrt{\frac{\bar{D}t}{\pi}} \tag{2}$$

where Θ_i = initial water content, Θ_0 = air-dryness value of water content, t = time, π = constant, and D = mean weighted diffusivity, computed by

$$\bar{D} = \frac{1.85}{\theta_i - \theta_0}\int_{\theta_0}^{\theta_i} D(\theta)(\theta - \theta_0)^{0.85}\, d\theta$$

It was assumed that (1) evaporativity is infinite, (2) during drying the soil, moisture diffusivity D changes exponentially with the water content (Gardner and Mayhugh, 1958), (3) the water content at the surface is instantaneously brought to the water content in equilibrium with the atmosphere, (4) the water content is uniform, and (5) at the boundry of a semifinite slab, the water content is constant and the initial and boundry conditions are

$$\theta = \begin{cases} \theta_i, & x \geqslant 0,\ t = 0 \\ \theta_0, & x = 0,\ t > 0 \end{cases}$$

Gardner and Hillel (1962) assumed constant diffusivity with depth and obtained the following solution for a finite system:

$$E = \frac{D(\bar{\theta})W\pi^2}{4L^2} \tag{3}$$

where E is the evaporation rate and Θ is the average water content obtained by dividing the total amount of water W by the depth of wetting L.

For these assumptions with regard to soil and environment Gardner (1959) and Gardner and Hillel (1962) found that the CE under higher E_0 at any time was higher than that under lower E_0. But these assumptions of infinite E_0, uniform moisture wetness, and neglect of gravity seem far from the real-world situation. The gravity term can be neglected for fine-textured soils, but not for coarse-textured ones because with gravity water gets redistributed more rapidly into deeper layers and becomes less accessible for evaporation from the soil surface. The effect of gravity is of greater consequence when E_0 is high and the soil texture is coarse, as it significantly influences the evaporation loss (Jalota and Prihar, 1985).

Recently Jalota and Prihar (1987) incorporated the gravity term in the flow equation, Equation (4), and solved it numerically in two ways,

$$\frac{\partial \theta}{\partial t} = \frac{\partial}{\partial z}\left[D(\theta)\,\frac{\partial \theta}{\partial z} - K(\theta)\right] \tag{4}$$

In one case the evaporation rate from the soil was equated with the flux density through the (shallow) plane at 2.5 cm. In the second procedure it was assumed that drying of the surface layer integrates the effects of E_0, initial wetness, and the $D(\Theta)$ or $K(\Theta)$ relations of the soil. The equation was solved by the finite-difference technique for the initial and boundary conditions as follows:

$$\theta(Z, 0) = \theta_0(Z) \quad \text{(initial condition)}$$
$$\theta(\Delta Z_i, t) = \theta_i(t) \quad \text{(upper boundary condition)}$$
$$\left(\frac{\partial \theta}{\partial Z}\right)_{Z=L} = 0 \quad \text{(lower boundary condition)}$$

The observed CE agreed well with that calculated by both procedures in sandy loam soil (Figure 1). The predictions by these numerical solutions of the flow equations and the experimental results of Jalota and Prihar (1986) as shown in Figure 2 clearly indicate that the effect of E_0 on CE depends on the soil type and the initial wetness of the soil. When evaporation from slow draining soils, in which the water transmission coefficient declines gradually with the decrease in water content, commenced immediately after wetting, the CE at any time up to 30 days remained higher under higher E_0, than under lower E_0. In soils of low retentivity or where the soil surface water content was lowered by redistribution before commencement of evaporation, CE under higher E_0 fell below that under lower E_0 after a certain period. Further, the evaporation rate remained higher under higher E_0 than under low E_0 as long as the decrease in soil water diffusivity under higher E_0 was accompanied by a large increase in hydraulic gradient. When the product of diffusivity and hydraulic gradient under higher E_0 became lower than that under lower E_0, the evaporation rates under higher and lower E_0 reversed. When this continues for a sufficiently long time, CE under higher E_0 lags behind that under low E_0 (Jalota and Prihar, 1987). Hanks et al. (1967) reported that for comparable E_0, drying by wind caused higher CE than drying by radiation. Under radiation conditions the warming of the soil surface raises the vapor pressure and causes thermally induced downward flow, which counters the upward water flow on account of the moisture gradient.

Jalota and Prihar (1986) also verified an earlier conclusion that evaporation rates during the falling-rate stage are insensitive to E_0 (Gardner, 1959). They found that evaporation did respond to E_0 during the falling-rate stage in silt loam and sandy loam but not in loamy sand soil. The magnitude and duration of the response of evaporation to E_0 were greater in finer textured than in medium textured soils and absent in coarse soils. Comparison of observed evaporation losses (Jalota and Prihar, 1986) with those predicted by different procedures showed that the procedure of Gardner (1959) gave a satisfactory prediction for loamy sand. Procedures of Gardner and Hillel (1962) and Jalota and Prihar (1987) were found suitable for sandy loam and silt loam under both E_0's.

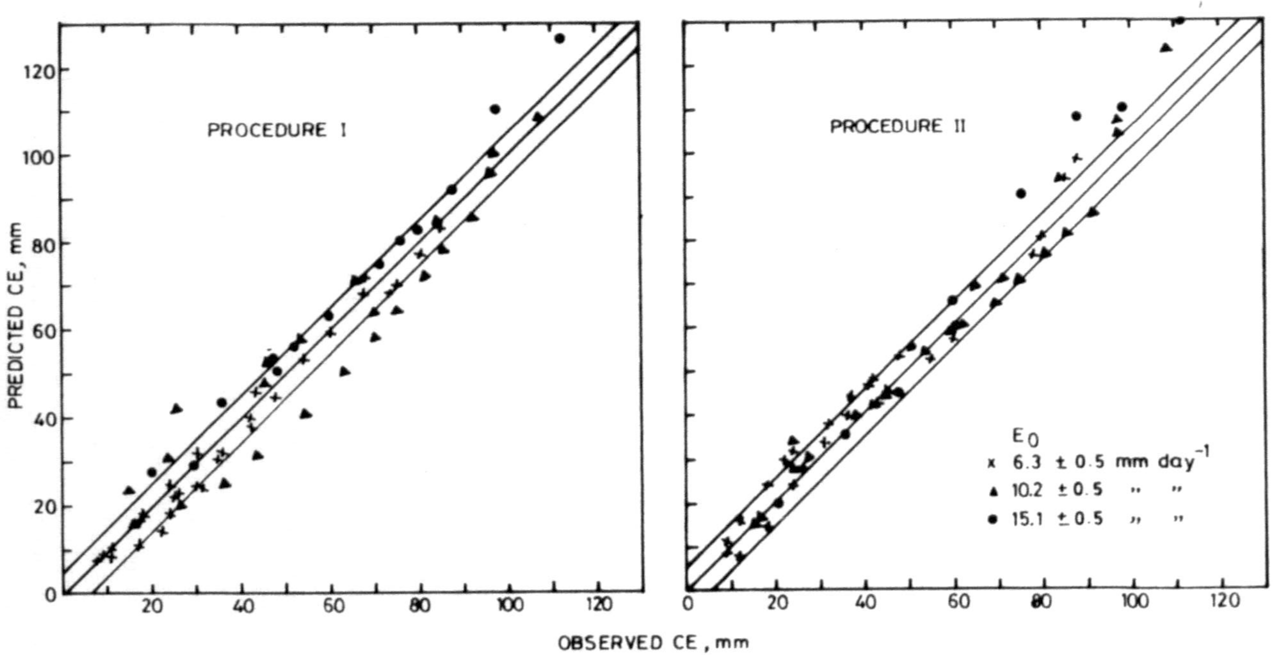

Figure 1. Comparison of observed and predicted evaporation by two procedures in sandy loam soil. [From Jalota and Prihar (1987).]

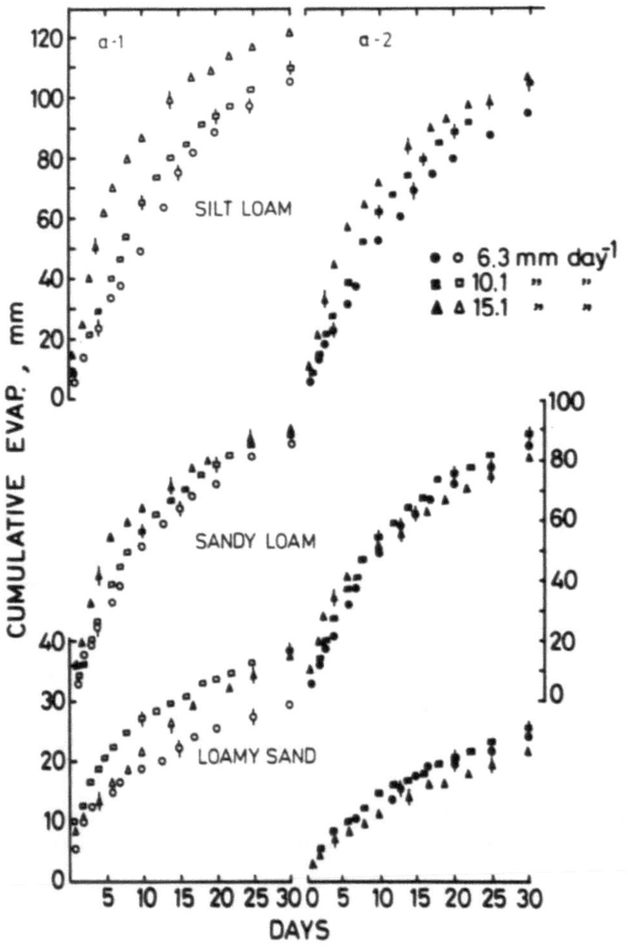

Figure 2. Cumulative evaporation from silt loam, sandy loam, and loamy sand immediately after wetting (*a*-1, hollow symbols) and with 2-day redistribution time (*a*-2, solid symbols) for low (○, ●), medium (□, ■), and high (△, ▲) evaporativities. [From Jalota and Prihar (1986).]

Rowse's (1975) procedure underpredicted evaporation in all three soils and two E_0's (Figure 3). Inferentially, the assumptions made by Gardner (1959) to achieve the exact solution of the flow equation are valid for coarse-textured soils alone, which dry quickly at the surface. Evidently these assumptions did not hold for silt loam and sandy loam. Jackson et al. (1976) showed that during the initial period of the falling-rate stage the surface is not completely dried and the wet part evaporates at a rate dictated by E_0 in addition to the transmission properties of the soil. Thus while predicting the evaporation rates during the falling-rate stage from medium and finer textured soils, it is important to account for the effect of E_0 directly (such as by involving the drying pattern of the surface layer), as was done by Jalota and Prihar (1987), or indirectly (including surface water

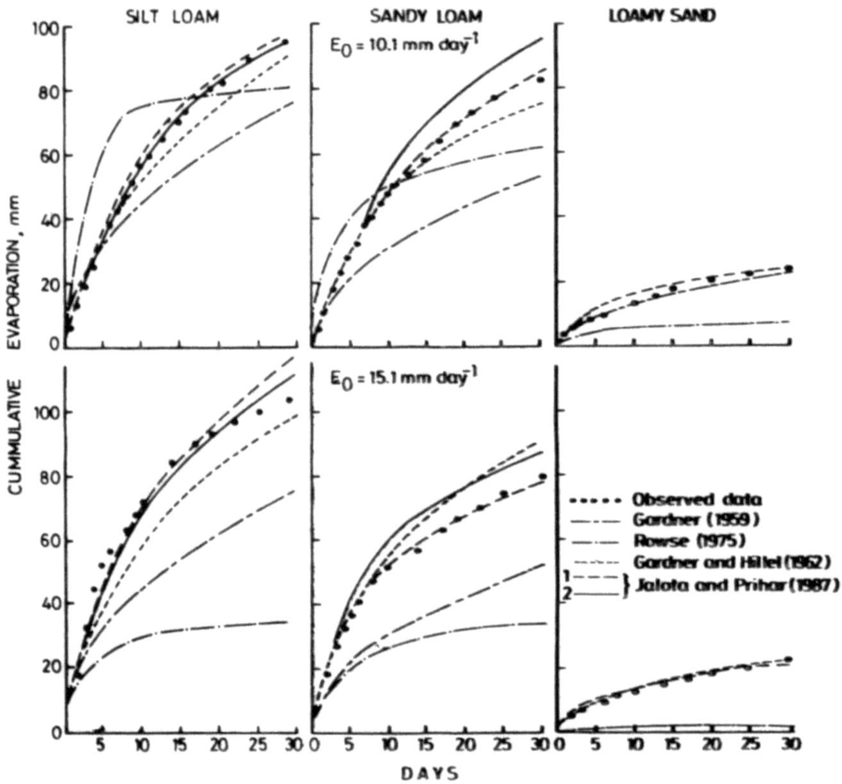

Figure 3. Comparison of observed evaporation during falling-rate stage and predicted by different procedures in silt loam, sandy loam, and loamy sand for two evaporativites.

content changes in the average water content of the wetted length), as was done by Gardner and Hillel (1962).

Several researchers have simplified the empirical relations between time and CE obtained from the analytical flow equation (Gardner, 1959), controlled drying of the laboratory columns (Rose, 1969; Klute et al., 1965), and field lysimeters (Black et al., 1969). The CE was generally reported to be a linear function of the square root of time during the falling-rate stage. Black et al. (1969) predicted CE through a simple linear relation involving $t^{0.5}$ from sandy soil under low E_0. However, Gardner and Gardner (1969), working with a fine-textured soil, found that CE deviated from a linear relation involving $t^{0.5}$. Kijne (1973) considered the evaporation losses from bare 13-cm diameter and 35-cm-deep soil columns of fine sandy loam and clay loam soils during all three stages of evaporation and found that CE during the first 4 days after drying started was best approximated by a linear function of $t^{0.67}$ and during the next 15 to 20 days by a linear function of $t^{0.40}$. More recently, Jalota (1984) also found the exponents of time to vary with soil type and E_0 (Table 1). While fitting of the

Table 1. Cumulative evaporation (CE) versus time (t) relations and periods for which these hold for different soils under two evaporativities

Soil type	Evaporativity E_0 (mm/day)	Best-fit equation of CE[a]	Period for which relation holds (days)	r^{2b}
Silt loam	15.1 ± 2.66	$14.1t^{0.72}$	0–7	
		$24.2t^{0.43}$	7–15	
		$30.13t^{0.36}$	15–30	
		$-3.12 + 19.68t^{0.5}$	0–30	0.932
	4.6 ± 1.08	$5.3t^{0.90}$	0–15	
		$20.29t^{0.36}$	15–30	
		$-9.88 + 15.63t^{0.5}$	0–30	0.975
Sandy loam	15.1 ± 2.66	$12.84t^{0.71}$	0–7	
		$27.62t^{0.29}$	7–15	
		$30.07t^{0.26}$	15–30	
		$11.37 + 11.92t^{0.5}$	0–30	0.925
	4.6 ± 1.08	$-6.61 + 11.06t^{0.5}$	0–30	0.994
Loamy sand	15.1 ± 2.66	$-0.0073 + 5.772t^{0.5}$	0–30	0.995
	4.6 ± 1.08	$-1.125 + 5.67t^{0.5}$	0–30	0.829

From Jalota (1984)

[a]CE is in millimeters and t in days.

[b]r^2 = coefficient of determination.

$t^{0.5}$ relation to CE a negative intercept was obtained, which became more negative with the fineness of the soil texture. The regression coefficient or slope of CE versus $t^{0.5}$ increased with the fineness of the texture and E_0. To account for these additional observations Jalota et al. (1988) modified the square root of time relation, including the constant-rate-stage evaporation as follows:

$$\begin{aligned} CE &= E_0 t, & 0 < t < t_f \\ CE &= Kt^{0.5} - \frac{K^2}{4E_0}, & t > t_f \end{aligned} \tag{5}$$

where E_0 = potential evaporation rate, K = regression coefficient of CE versus $t^{0.5}$, and tf = duration in days of the constant-rate stage and is related to K and E_0 as follows:

$$t_f = \frac{K^2}{4E_0^2}.$$

B. Effect of Soil Type

Under a given E_0, the soil type affects the CE through its retention and transmission properties. In the wet range the transmission rate is higher in coarse than in fine-textured soil, while the reverse is true in the dry range (Wind, 1961).

Therefore when evaporation commences from a nearly saturated soil, having moisture at very low tension, evaporation is initially higher from coarse soil (Penman, 1941; Prihar et al., 1968), and the trend reverses after some time (Gill et al., 1977). On the other hand when evaporation starts from a soil with initial moisture content at or near field capacity, CE from a fine soil will remain higher than that from a coarse-textured soil (Jalota and Prihar, 1986).

C. Effect of Redistribution Time

Redistribution of water allowed prior to the commencement of evaporation or occurring simultaneously with evaporation reduces evaporation losses. This is so because it decreases the water content and hence the overall gradient and soil water diffusivity in the upper zone subject to evaporation. Reduction in CE with postirrigation redistribution under laboratory (Staple, 1976) and field (Jaggi et al., 1978) conditions and with redistribution during water application at low rates (Bresler et al., 1969) has been reported in the literature. These studies were conducted with a single soil subjected to a fixed E_0. Recently Jalota and Prihar (1986) showed that the effect of redistribution time on evaporation reduction was influenced by the texture [or the $K(\Theta)$ relation] of the soil and the prevailing E_0. Redistribution brought about greater reduction in evaporation on coarse than on fine-textured soils (Figure 2), which is attributed to faster drainage and lower retention of water in the surface layers of the coarse soil. In coarse soils water moves into lower layers and becomes less accessible to evaporation losses. Lower water contents in surface layers result in earlier formation of the dry layer, especially under high E_0, which reduces further evaporation losses.

IV. Dynamics of Evaporation from Tilled Soil

The loosening of surface soil by tillage brings about significant changes in its energy reflection and water transmission properties because of the way evaporation from tilled soil deviates from that from untilled soil. Linden (1982) reported that the loosening of soil increases its surface roughness, which reduces albedo and increases potential evaporation (Allmaras et al., 1972; Bowers and Hanks, 1965; Cary and Evans, 1975; Idso et al., 1975; Van Bavel and Hillel, 1976) by concentrating heat in the surface layers. Roughness also increases the surface areas of soil exposed to the atmosphere and allows greater penetration of wind (Ojeniyi and Dexter, 1979), both of which are conducive to an increase in evaporation. The hydraulic properties of soil are also altered by tillage. Bulk density and pore size changes associated with loosening cause soil suction to change at a given volume water fraction and, thus, change hydraulic conductivity (Box and Taylor, 1962; Archer and Smith, 1972). A decrease in bulk density increased retention at low suction and decreased it at high suction (Laliberte and Brooks, 1967). Similarly a decrease in bulk density increased hydraulic conductivity at low suction and decreased it at high suction.

Decreased albedo, an increased surface area exposed to the atmosphere, and greater penetration of the wind into the tillage-induced loose and rough surface layer increases the initial rate of evaporation from the tilled soil. This rapid loss of water from the upper portion of this layer is replenished more slowly in the tilled compared with the untilled soil on account of decreased hydraulic conductivity of the loosened layer. Hence the top few centimeters of the tilled layer dry out more rapidly and intensively than the corresponding layer of the untilled soil. This dry layer acts as a barrier to both liquid and vapor flow to the surface and reduces further evaporation from the tilled compared with the untilled soil. This is so because the untilled soil dries out more gradually and permits liquid flow from the interior of the soil to the surface for a longer period. Breaking down the capillaries by tillage retards the hydraulic conductivity or diffusivity of the profile, particularly of the surface zone, and the potential gradient at the surface. But an important phenomenon occurs at the base of the tilled layer, that is, the lower portion of the tilled layer may permit upward capillary flow to some distance above the interface, depending on the fineness of clods, the water content of the tilled layer, and the type of soil. Gill and Prihar (1988) presented evidence of liquid-water flow from the untilled soil below to seeds placed some distance above it in the (simulated) tilled layer. When the tilled layer was separated from the untilled soil below by a thin layer of water-proofed aggregates, which permitted only vapor flow, water flow to the seeds was reduced. This implied that in the absence of water-proofed layers, the liquid water moved to the seeds through the loose soil. Evidently, therefore, the phase change from liquid water to vapor in the tilled soil does not occur at the interface of tilled and untilled soil, but at various loci within the tilled layer. Nevertheless, formation of a dry layer results in (1) pushing the locus of conversion of liquid water to vapor some (variable) distance below the surface; and (2) lowering the temperature at the interface of tilled and untilled layers and, thus, reducing the saturated vapor pressure at that plane. It also resists the flow of vapor from soil to atmosphere and conducts liquid water much more slowly than the undisturbed soil below.

With time the surface layer of the untilled soil also dries out and the locus of phase change moves deeper into the soil. The dry layer that develops at the surface of the undisturbed soil is less porous than that created by tillage. When the drying front moves deeper into the untilled soil, the less porous dry surface layer permits less vapor flow compared with an equal depth of the more porous layer of the tilled soil (Acharya and Prihar, 1969). Consequently the rate of evaporation loss from the tilled soil begins to exceed that from the untilled soil. The time course of evaporation flux from tilled soil normalized against that from untilled soil is shown schematically in Figure 4. For a short time after tillage the evaporation rate from tilled soil exceeds that from untilled soil. With rapid formation of a dry layer because of accelerated drying of the tilled surface, evaporation from the tilled soil lags behind that from the untilled soil. Eventually, when a dry layer also forms at the surface of untilled soil, evaporation from the tilled soil again exceeds that from the untilled soil. The duration and magnitude of this increase or decrease in evaporation rates in tilled vis-à-vis

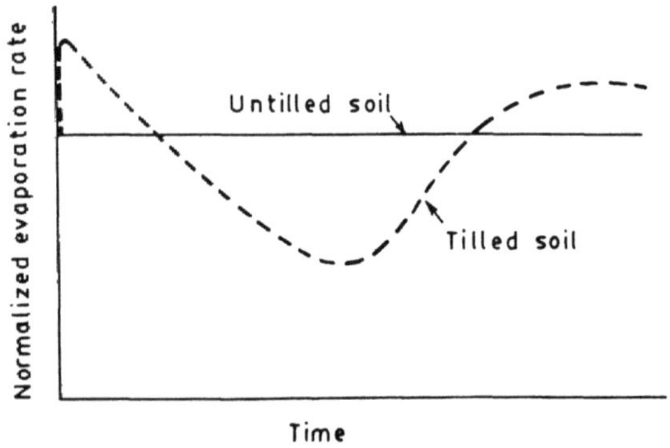

Figure 4. Schematic representation of evaporation rate from tilled soil relative to that from untilled soil.

untilled soil, which determines the course for evaporation reduction with tillage, have been found to depend on the soil type, evaporativity, time and type of tillage, and their interactions.

V. Evporation Reduction with Tillage

Shallow tillage is often advocated for reducing evaporation losses from the soil, but its role in doing so is not unequivocally understood. The concept of tillage for moisture conservation originated from the famous laboratory experiments of King (1907) carried out in the United States in Wisconsin toward the end of the nineteenth century. These experiments showed that soil mulch, created by loosening a shallow layer on top of 120-cm-tall columns, kept in contact with free water at the bottom, caused a large reduction in evaporation losses. In a pot culture study in Utah 2.5-cm cultivation reduced evaporation losses by 34, 13, and 63% in sand, sanpete clay, and clay, respectively (Widstoe, 1909). Similar results had been obtained by Woolney and his school in Germany (Eser, 1884). Based on these and other similar results, dust mulch was advocated for conserving soil moisture against evaporation. These findings were not, however, upheld by field experiments (Call and Sewell, 1917; Veihmeyer, 1927) in the absence of a water table, where the weeds were kept down by surface scraping. It was argued that in the absense of a steady upward supply from a water table most water that can be lost by evaporation was already lost by the time the soil permitted the creation of soil mulch by tillage. Jacks et al. (1955) concluded from the then available literature on the subject that the creation of soil mulch was not helpful in reducing evaporation except by killing weeds. About a decade later Pendelton (1966) also stated that the theory of dust mulch for moisture conservation had long been abandoned by agronomists. But there were reports in the literature which showed that soil mulch did help in reducing evaporation. For example, in a 4-year study Singh and Nijhawan (1944) found evidence of moisture conservation by soil mulch created in situ.

Benoit and Kirkham (1963) reported reduced evaporation with soil mulch applied extraneously to soil columns wetted to field capacity. But their experimental setup did not simulate the in situ soil mulch created by tillage. Prihar and VanDoren (1967) studied evaporation losses from 15-cm-diameter, 40-cm-deep "baby lysimeters" filled with silt loam soil and placed between rows of growing corn with tops flush with the soil surface. Stirring the surface 5-cm layer with a garden hoe after each rain reduced short-term (3- to 5-day periods) evaporation losses to almost 50%. Aujla and Cheema (1983) reported that one shallow cultivation of a bare loamy sand field in September, after cessation of the monsoon rains at Ludhiana, India, reduced evaporation loss by 20 mm in 46 days, compared with no cultivation. Recently Minhas et al. (1986) studied evaporation loss as affected by tillage, employing the zero-flux-plane method of Ehlers and Van Der Ploeg (1976), from sufficiently wetted plots of silt loam soil either tilled 5 cm deep or left untilled. Evaporation rates for 18 days after tillage averaged 1.67 mm per day in the tilled and 2.16 mm per day in the untilled plots, while E_0 averaged 3.17 mm per day. Under an average E_0 of 8.47 mm per day the tilled and untilled plots lost 3.80 and 4.26 mm water per day, respectively.

A number of laboratory (Prihar et al., 1968; Willis and Bond, 1971; Gill et al., 1977; Gill and Prihar, 1983; Prihar and Jalota, 1988) and field (Papendick et al., 1973; Jalota and Prihar, 1979) experiments have shown that the effect of soil mulch on evaporation reduction is governed by soil type, initial wetness, evaporativity, time of creation of soil mulch, and thickness and clod size distribution of the mulch. Moreover, the magnitude of the benefit is time variant.

A. Effect of Soil Type

In soils that retain more water and where the hydraulic conductivity decreases gradually with the decrease in water content, the surface soil remains moist for long periods because the high rate of upward flow continues to replenish the water lost from the upper layers. In such cases the locus of evaporation can be shifted effectively some distance below the soil surface by loosening the upper several centimeters layer by tillage to retard upward flow. This would hasten the development of a dry surface layer and, thus, reduce evaporation from the tilled compared with the untilled soil. On the contrary, in coarse soils of low retentivity a dry layer forms at the surface irrespective of tillage because the hydraulic conductivity decreases sharply with a decrease in surface layer water content and the upward flow is inadequate to replenish the water lost from the upper layer. In such soils the advantage of tillage to break the capillary action is considerably smaller than in fine-textured soils. The limited (controlled) experiments involving more than one soil (Gill et al., 1977; Prihar et al., 1968) corroborate this line of reasoning. In another study 30-day evaporation rates from tilled and untilled soils averaged 1.98 and 3.13 mm per day in silt loam and 1.41 and 2.13 mm per day in sandy loam soil, respectively (Jalota and Prihar, unpublished). Obviously, tillage was more effective in water conservation on fine soils than on coarse-textured soils.

B. Effect of Evaporativity

Under otherwise comparable conditions of soil and initial wetness, high E_0 enhances the rate of water loss from the soil surface layers. This reduces the water content and hydraulic conductivity of the upper soil, but at the same time it increases the hydraulic gradient. The latter tends to sustain the upward flow for some time. Eventually, however, the flow from lower layers fails to replenish the water loss from the surface soil and surface layers dry (Papendick et al., 1973). The thickness of the dry layer that develops at the surface is just sufficient to balance the upward rate of flow. If a layer of higher resistance is created, such as through tillage, the evaporation rate will decrease. Therefore shallow tillage, which hastens the formation of a dry layer, results in substantial reduction in evaporation over shorter periods. But with continued drying, a dry layer also develops at the surface of untilled soil and the evaporation rate is reduced, often below that from the tilled soil. Therefore under high E_0, evaporation reduction peaks after a certain length of time and reduces thereafter. After a sufficiently long time the CE from tilled soil may equal or exceed that from untilled soil (Wiese and Army, 1958).

When the E_0 is low, the upward flow replenishes the small amount of loss and keeps the soil surface moist for a long time. Penman (1941) has shown that a steady demand on soil water supply could be met for a long time. Tillage reduces the upward flow by breaking the capillaries and, thus, reduces the evaporation rate compared with the untilled soil. Since the dry layer on the untilled soil does not develop for a long time, evaporation reduction with tillage under low E_0 continues to increase for much longer periods compared with that under high E_0. Jalota and Prihar (1989) observed that evaporation reduction with tillage of sandy loam soil peaked at 10 days after tillage under E_0 of 15.1 mm per day, but continued to build up for 30 days under E_0 of 4.6 mm per day.

C. Interactive Effects of Soil Type and Evaporativity

The soil type has been observed to interact strongly with evaporativity for moisture conservation by tillage (Prihar et al., 1968; Gill et al., 1977; Jalota, 1984). Soil mulch created by tillage was found beneficial for moisture conservation on all soils under low E_0. But under high E_0 the benefit of tillage was confined to fine-textured soils only. In fact, various combinations of E_0 and soil type yield a range of longevity and magnitude of evaporation reduction by tillage. The two extremes are represented by a fine-textured soil under low E_0 and a coarse soil under high E_0. When the soil is coarse and E_0 is high, a rapid decrease in surface water content is accompanied by a sharp decrease in hydraulic conductivity, and a dry layer is formed rapidly whether the soil is stirred or not. Under such conditions, tillage may hasten the formation of a dry layer only

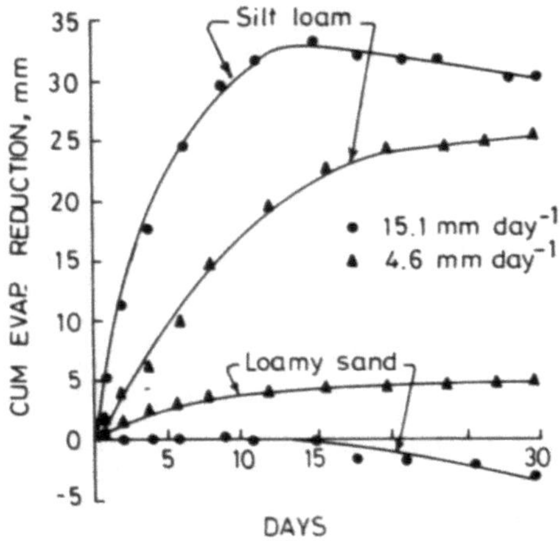

Figure 5. Evaporation reduction by tillage as a function of time in silt loam and loamy sand for two evaporativities.

slightly and, hence, offers little benefit for moisture conservation (Prihar et al., 1968). Interestingly, tillage of loamy sand was observed to have a negative effect on moisture conservation under high E_0 (Jalota, 1984). However, where E_0 is low, the liquid flow, even in coarse-textured soils, continues for some time and the hastening of the formation of a dry layer through tillage helps in reducing evaporation to some extent (Figure 5). On fine-textured soils under low E_0 cumulative reduction in evaporation continued to increase for longer periods as the formation of a dry surface layer in the untilled soil took longer.

D. Effect of Time of Tillage

Hanks and Gardner (1965) demonstrated that changes in diffusivity in the wet range have a large influence on the CE from the soil. Since the $K(\Theta)$ and $D(\Theta)$ functions of fine-textured soil differ a great deal from those of coarse-textured soil, it is to be expected that the soil type and wetness at which $D(\Theta)$ change is caused by tillage will influence the course of evaporation. Willis and Bond (1971) tilled sandy loam soil columns exposed to an E_0 of 9.6 mm per day at 1, 4, 7, and 18 days after wetting and found that earliest tillage gave the maximum reduction in evaporation. They also observed that when tillage was delayed for 18 days, 7.5-cm-deep tillage tended to be more beneficial than 2.5-cm-deep tillage. The effect of the time of tillage also varied with soil type, tilth, and E_0, as shown by Gill et al. (1977). They reported that under low E_0, tillage of the sandy loam soil 5 and 11 days after wetting caused greater evaporation reduction than at 18 and 42 days after wetting; but under high E_0 the earliest tillage was most effective. Interestingly, in the silty clay loam the second time of tillage (11 days after wetting) was most effective (Table 2). It is postulated that when fine-textured soil is tilled in wet condition under high E_0, it tends to reconsolidate and

Table 2. Cumulative evaporation loss (mm) during 67 days under low E_0 and 52 days under high E_0 from silty clay loam (Sicl) and sandy loam (sl) soils in relation to time of tillage[a]

Time of tillage (days after wetting)	Low evaporativity (2.0 mm day)		High evaporativity (9.8 mm day)	
	Sicl	Sl	Sicl	Sl
5	60.6a*	56.8a	107.5c	85.1a
11	56.7a	58.1a	94.5a	86.4ab
18	60.2a	63.2b	102.8ab	88.4ab
42–43	69.3b	73.6c	104.2bc	88.5ab
Untilled	83.6c	77.6d	103.7abc	90.0b

From Gill et al. (1977). Reproduced from the *J. Indian Soc. Soil Science*, 25:360–366 (1977) with permission from the Indian Society of Soil Science.

[a]Any two means not followed by the same letters are significantly different at the 5% level.

reestablish the capillarity. Further, coarse tilth (MWD 45 to 54 mm) was found more effective in reducing evaporation loss when tillage was performed early, while fine tilth (MWD 7 to 13 mm) was more effective with delayed tillage.

Thus for a given time of tillage, the characteristics of the tilled layer, such as porosity, clod size distribution, and depth of tillage, influence the extent and time course of evaporation loss.

E. Effect of Characteristics of Tilled Layer

1. Air Porosity

The air porosity of mulch has a direct relation to evaporation loss (Hanks, 1958; Acharya and Prihar, 1969). With an increase in the porosity fraction of the tilled layer from 0.5 to 0.64, evaporation increased by 25% (Allmaras, 1967). The reduction in porosity increases the diffusion resistance to vapor flow by making the flow path more tortuous. On this basis, compaction of the dry tilled layer is advocated to lower evaporation losses. But another effect of reduced porosity is to increase the thermal conductivity of the tilled layer (Papendick et al., 1973), which would tend to increase evaporation. On the other end of the scale, a decrease in thermal conductivity with an increase in porosity would enhance the vapor exchange by increasing the turbulence of air (Hanks and Woodruff, 1958; Acharya and Prihar, 1969). Therefore to achieve maximum evaporation reduction, we must look for that threshold porosity of the tilled layer which reduces thermal conductivity without any increase in the turbulent flow of air.

Since porosity of the tilled layers is a function of clod size distribution, a number of researchers have attempted to define the optimum clod size range for maximum evaporation reduction (Table 3). The inconsistency in the optimal size of aggregate was due to the variable conditions of experimentation. Heinonen

Table 3. Optimum clod sizes for evaporation reduction as given by different workers

Reference	Clod-size distribution diameter (mm)
Wadham (1944)	<10
Ramacharlu (1957)	0.8
Holmes et al. (1960)	2.5
Scotter and Raats (1969)	0.3–0.6
Hillel and Hadas (1972)	0.5
Heinonen (1985)	0.5–2.0

(1985) showed schematically the main processes of water loss associated with the size of aggregates (Figure 6). The first peak of the evaporation rate occurred for aggregates of 0.05 mm due to high hydraulic conductivity of the aggregates. Increasing the aggregate size from 0.5 to 5.0 mm increased the air-filled porosity and slightly increased loss by gas diffusion. Wind-induced turbulence in soil pores started when the aggregate size increased to 3 to 5 mm and became dominant for aggregate sizes exceeding 10 mm. The maximum drying rate was obtained in 5- to 10-mm-diameter and larger clods. The surface area became a limiting factor for evaporation reduction, especially if hydraulic conductivity within the clod was low. In fact, the optimal clod size would depend on the properties of the soil and wind gustiness.

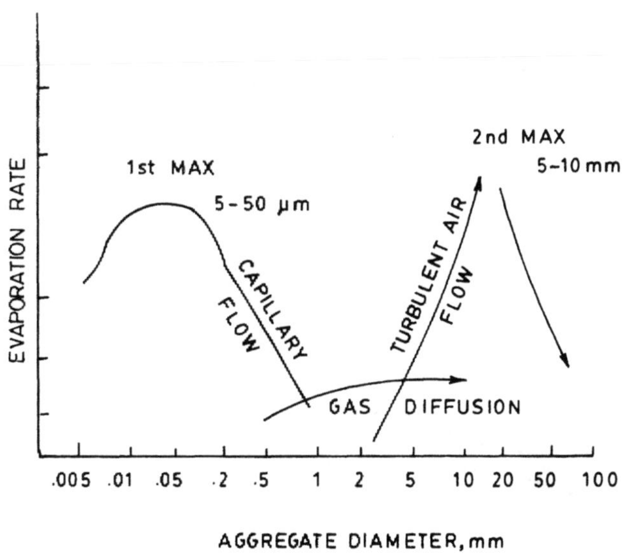

Figure 6. Schematic representation of evaporation rate and water transport process in relation to aggregate size. [From Heinonen (1985).]

2. Depth of Tillage

Using surfactant-treated soil mulch Hanks (1958), Acharya and Prihar (1969), and Hillel et al. (1975) showed that evaporation loss was inversely related to mulch thickness. When the depth was less than a certain threshold value, evaporation was limited by external conditions rather than by mulch (Gardner and Fireman, 1958). Deeper tillage reduces evaporation to a greater extent by providing a stronger barrier to vapor diffusion (Willis and Bond, 1971; Gill and Prihar, 1983).

Actually, the tilled layer starts drying from the top and the drying front moves downward into the soil until the dry layer is sufficiently thick to reduce the loss rate such that it is balanced by unsaturated upward flow dictated by soil and climatic conditions. Jalota (1984) computed the thickness of an (absolutely) dry layer that permitted the same evaporation (vapor) loss as the tillage-induced soil mulch at a given point of time. The thickness of the dry layer was found to increase with time at least until 30 days after tillage under low E_0 of $4.9 + 1.1$ mm per day in silt loam, sandy loam, and loamy sand soils. But under high E_0 of 15.3 ± 2.6 mm per day, it attained the maximum value of 1.2 cm after 28 days in sandy loam (Figure 7). These results and those of Papendick et al. (1973) showed

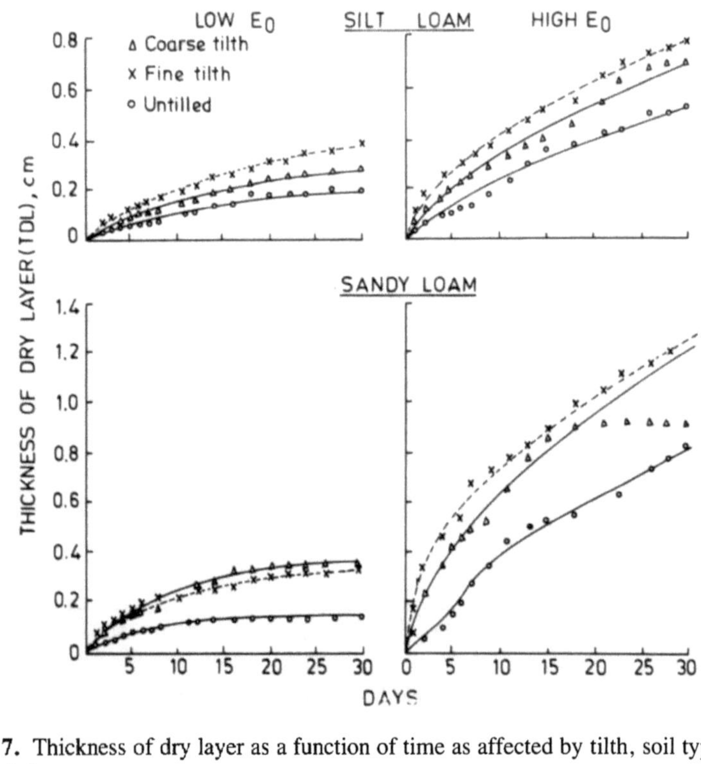

Figure 7. Thickness of dry layer as a function of time as affected by tilth, soil type, and evaporativity.

that the formation of the dry layer is a dynamic process, and its rate of formation and its effective depth are governed by E_0, the transmission properties of the soil, and the roughness of the tilled layer. When the thickness of dry soil mulch was less than the depth of tillage, findings indicated that the water in the tilled layer did not move as vapor alone as is the case with completely dry soil. Possibly, the liquid flow from the undisturbed layer continues through or within the tilled layer, albeit at a lower rate than in untilled soil. Shallower tillage could possibly be as effective in reducing evaporation as 5-cm-deep tillage, provided the liquid flow from lower layers to the tilled zone is completely cut off through accelerated drying of the tilled layer with additional stirring. This statement was verified through experimentation, and the results are shown in Figure 8.

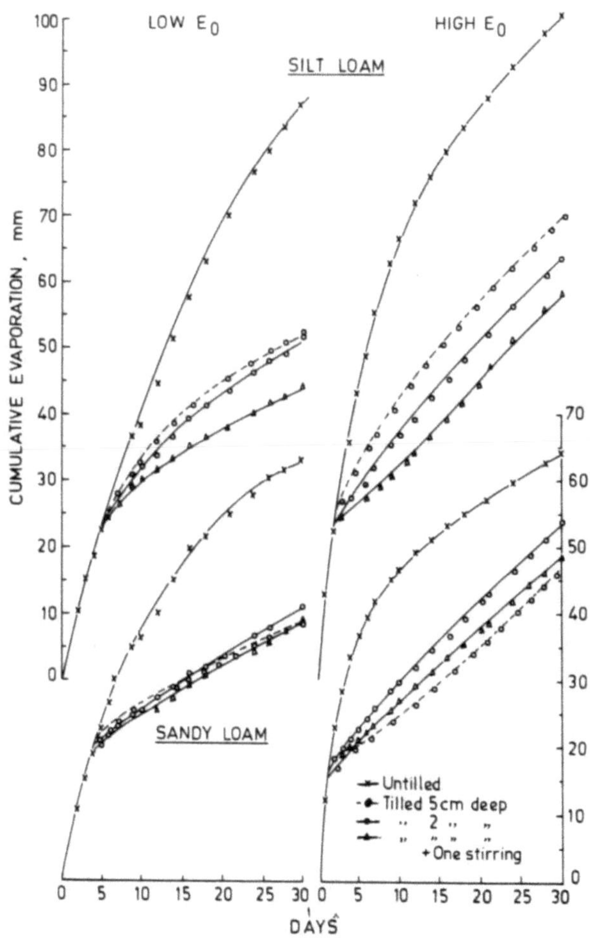

Figure 8. Time pattern of cumulative evaporation as affected by depth of tillage, soil type, and evaporativity.

3. Tillage Depth and Soil Type Interaction

The depth of tillage also showed an interaction with the soil type. Under high E_0, 2-cm-deep tillage was found more effective than 5-cm-deep tillage in the silt loam, while the reverse was true in the sandy loam. In the silt loam shallower loosening caused more rapid drying of the surface layer and was initially more effective in reducing evaporation than the deeper loosening (Figure 9). With the deeper (5-cm) tillage, the dry layer developed more slowly because the lower part of the tilled layer supplied water for evaporation. However, the dry layer thickened with time, and by the twelfth day a thicker layer had developed in the deeper tillage than in the shallower tillage. Consequently the evaporation rate with 5-cm-deep tillage became lower than that with 2-cm-deep tillage. In the sandy loam this phenomenon was absent. Apparently the stirred layer dried rapidly, irrespective of the depth of tillage. On account of its lower water

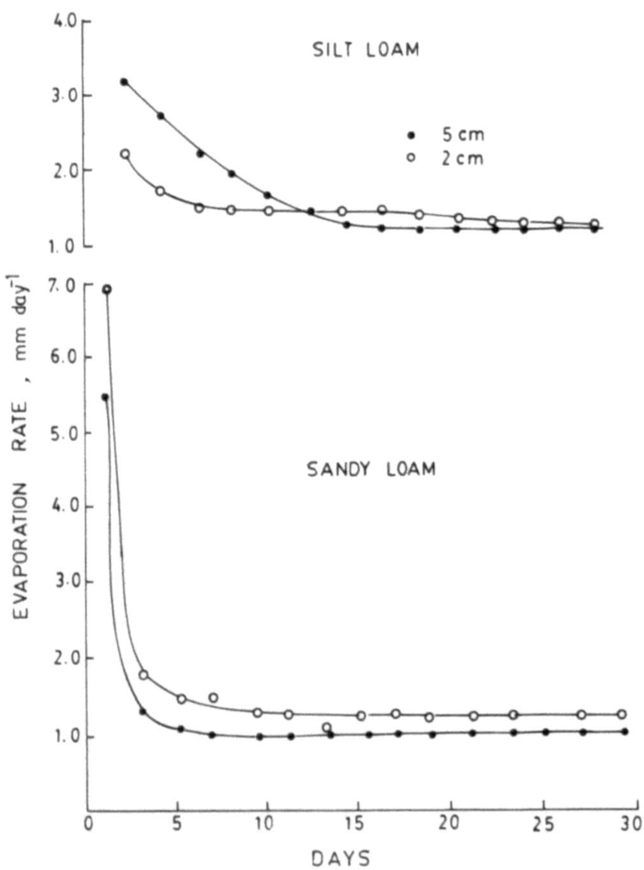

Figure 9. Time pattern of evaporation rate under high E_0 (15.3 mm/day) as affected by depth of tillage and soil type.

retentivity and lower diffusivity, the lower part of the tilled soil (in deeper tillage) did not transmit water fast enough to the upper part. Continuity of the upward liquid supply was also evident from the higher flux density through the junction of tilled and untilled soil than through the water-proofed (aggregate) layer of equal thickness (Figure 10). The vapor flux density, represented on the X axis of Figure 10, with water-proof aggregates of 5- and 2-cm depth was 0.91 and 1.39 mm per day in silt loam and 0.72 and 1.15 mm per day in sandy loam, respectively. With 5- and 2-cm-deep tillage, the ratios of total flux density to vapor flux density on the first day were 4 and 3 in silt loam and 5 and 2.4 in sandy loam. These ratios decreased with time and stabilized after 6 days at 1.5 for 2-cm-deep tillage in silt loam and at about 1.6 for both tillage depths in sandy loam. For 5-cm-deep tillage of silt loam the ratio dropped to 1.7 and stabilized after 20 days. The higher flux density across the tilled layer than through a layer of equal thickness of water-proof aggregates even after 30 days of tillage indicated that tillage did not completely cut off the capillary flow between tilled and untilled layers; it only retarded it by decreasing the particle contacts and changing the $D(\Theta)$ relation.

The evaporative flux has three components, vapor flux and liquid flux across

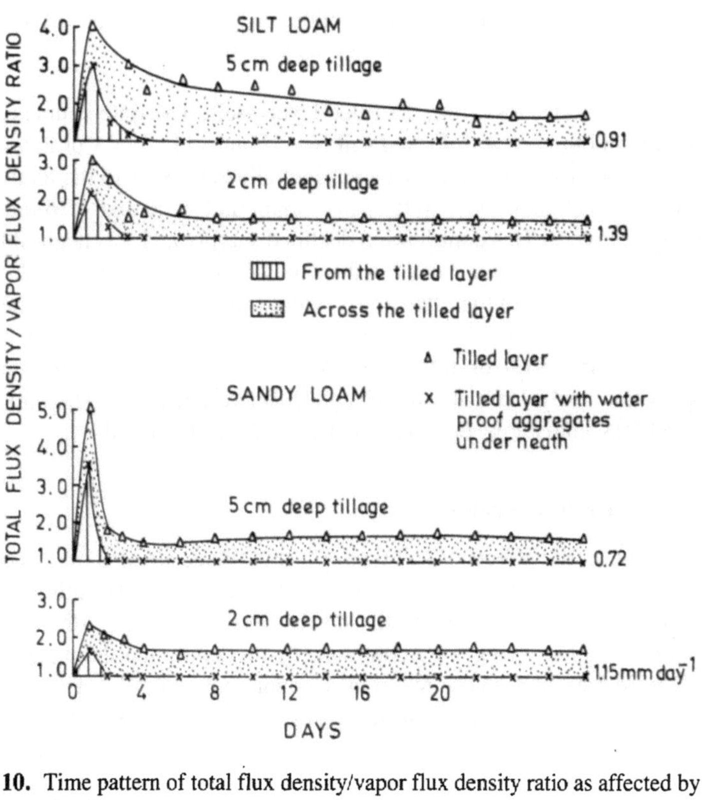

Figure 10. Time pattern of total flux density/vapor flux density ratio as affected by depth of tillage and soil type.

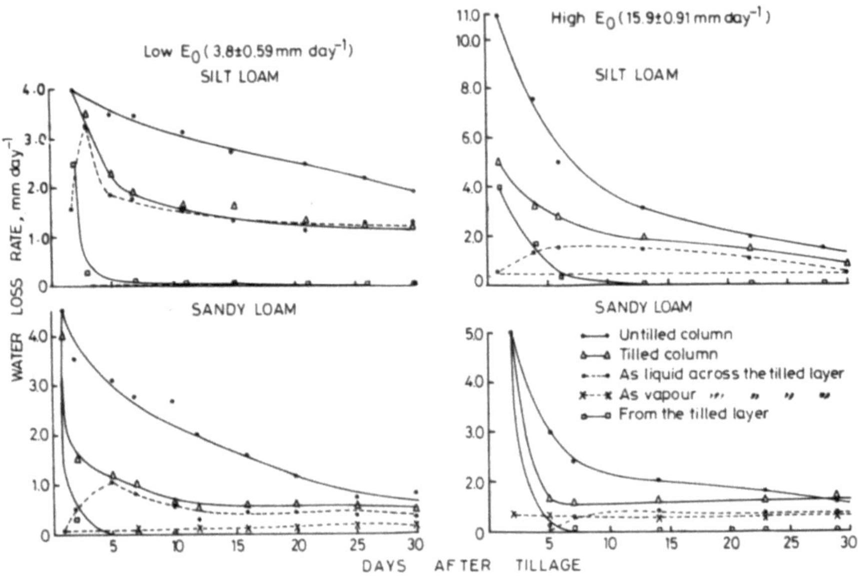

Figure 11. Time trend of water-loss rate of untilled and components of tilled columns of silt loam and sandy loam for low and high evaporativity.

the junction and water loss from the tilled layer itself. The time course of these fluxes was found to vary with E_0 and soil type (Figure 11). For 1 or 2 days after tillage water loss from the tilled zone contributed 62 and 88% to the total flux in the silt loam and 80 and 100% in the sandy loam under 3.6 and 15.6 mm of E_0 per day, respectively. This contribution declined sharply and was negligible after 12 days in the silt loam and 7 days in the sandy loam, irrespective of E_0. With decline in water loss from the tilled zone, liquid and vapor fluxes from the lower untilled soil started. Liquid flux in silt loam during the 30-day period after tillage constituted 72 and 36% of total loss under low and high E_0, respectively, whereas in the sandy loam it was 32 and 20%, respectively. The vapor flux contributed less under low E_0 than under high E_0 in both soils.

F. Drying Pattern of Profile

In addition to CE, the effects of soil type, E_0, and tillage are also reflected in the depth distribution of water in the soil profile. Gill et al. (1977), Gill and Prihar (1983), and Jalota (1984) reported that in the absence of tillage the soil water content profiles after drying were continuous with depth, and the water content gradients were greater in coarse-textured soils and under high E_0. These findings are in line with the theoretical considerations of Covey (1963) developed from the solution of the flow equation incorporating the effect of E_0, the length L of the soil column, the initial wetness θ_i, and the soil water diffusivity function ($D = A \exp B\theta$) in a single parameter G as

$$G = BE_0 \frac{L}{\theta_i}$$

He concluded that when G is small, resulting from a slow rate of drying, short columns, and a higher initial water content, the soil columns dry nearly uniformly with depth and present at all times a flat profile of moisture content. When G is large, the exposed surface dries much faster than the underlying material. The tilled layer above the interface dries out faster than the corresponding depth of untilled soil because of the reasons mentioned earlier. This phenomenon gets more pronounced with increasing E_0. The rate of advance of the isohydral front in the tilled zone is relatively rapid until it reaches the bottom of the tilled layer. After reaching the junction, its rate of advance slows down to a considerable extent, irrespective of E_0 (Gill and Prihar, 1983). However, the slow rate of advance was prolonged with deeper tillage and a decrease in E_0. Soil layers below the interface dry more slowly than the corresponding layers of untilled soil (Hillel and Hadas, 1972) because of the resistance of the interface to vapor flow. In the tilled soil more moisture is retained in the layers below the tilled zone (Sandoval and Benz, 1966). The depth of tillage also affects the water content profile. Shallow tillage (2 cm) caused a higher water content at a plane 3 cm below the surface, located in the 5-cm-deep tilled zone, irrespective of E_0 and soil type (Figure 12). The advantage was greater where 2-cm-deep tillage was stirred again. Jalota and Prihar (unpublished) observed that 30 days after tillage in silt loam under low E_0 (4.6 mm per day), the residual soil water content at a 3-cm plane averaged 9 and 15% with 5- and 2-cm-deep tillage and 18% when the 2-cm tilled layer was stirred again after 1 day. In the sandy loam, under the same

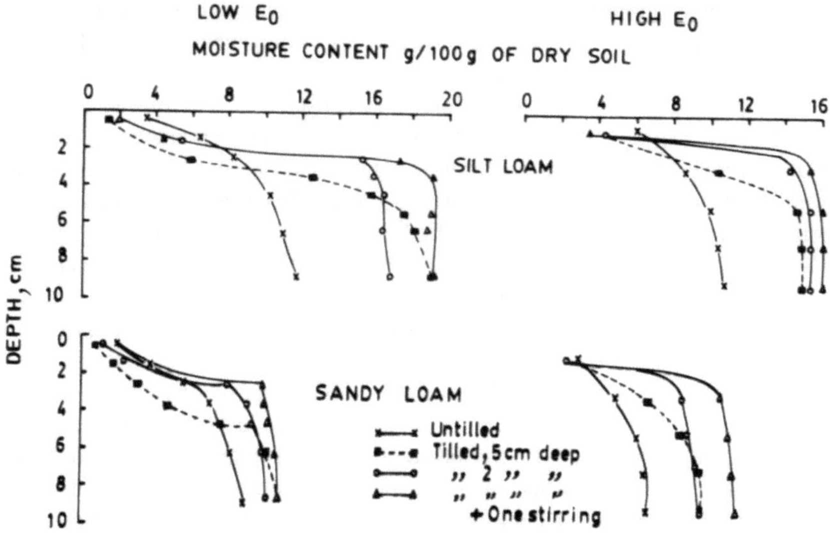

Figure 12. Near-surface moisture profiles at 30 days of drying as affected by depth and frequency of tillage, soil type, and evaporativity.

conditions, it was 3 and 8% with 5- and 2-cm-deep tillage and 10% for the 2-cm tilled layer stirred again. Under high E_0 (15.6 mm per day) the soil water content of this layer was 4 and 2% higher with 2-cm-deep tillage than with 5-cm tillage in silt loam and sandy loam soil, respectively.

G. Prediction of Tillage Effect on Evaporation

Methods of predicting evaporation from tilled soil range from a statistical approach to soil–water–atmosphere flow systems. Hanks (1958) and Hanks and Woodruff (1958) attempted to model the effect of tillage depth and porosity of the tilled layer on evaporation using surfactant-treated soil, assuming instant drying of the soil mass. They computed the vapor flux across the (simulated) tilled layer assuming the lower boundary of the layer as the locus of phase change. Evaporativity was either accounted for in the computation as vapor pressure at the surface (Hanks, 1958) or included as such (Acharya and Prihar, 1969; Budogoveskly, 1962). Likewise, Gill and Prihar (1983) assumed that the effect of cultivation was similar to that of intensely high E_0 because both conditions hasten the dry-layer formation. They modeled evaporation as a linear function of the square root of time. The depth of tillage and the effects of E_0 were reflected in the regression coefficients.

Evaporation has also been computed from moisture profiles at various times, predicted from the solution of the flow equation under isothermal conditions (Gardner, 1959; Hanks et al., 1969; Staple, 1971; Reynolds and Walker, 1984). Philip (1957) concluded from flow theory that temperature-induced moisture flow was relatively unimportant at low water contents and can be neglected. Thereafter, using arbitrary soil properties and weather patterns, Hillel and Talpaz (1977) simulated soil water dynamics for isothermnal conditions and predicted evaporation in variously constituted layers of nonhomogeneous profile. But these studies did not take into account interfacial impedance to water flow, which is very important in layered soils (Hillel and Hadas, 1972). However, later on Hillel et al. (1975) incorporated a diffusion resistance parameter of the soil mulch, which depended on the porosity, depth, and tortuosity of the soil, in the well-known combination equation of Van Bavel (1966) and simulated the dynamics of water storage in the profile. In this study they assumed that the value of E_0 and the evaporation rate from mulch-covered surface E_m are the same and heat flux is unaffected by the mulch. But as mentioned earlier, tillage alters the heat flux through changed roughness, which in turn changes the contact area of the soil with the atmosphere. Under such conditions drying becomes nonisothermal. Therefore the use of the isothermal model might be introducing an error in the estimation of evaporation. Hadas (1975) and Hanks et al. (1967) estimated that this error was as much as 20% for the homogeneous soil and will be more for the layered soil, especially when there are thermal fluctuations. Under such conditions vapor flow contributes significantly at intermediate water contents (Jackson et al., 1974). The contribution of vapor flow was taken into account by Hammel et al. (1981) while solving the flow equation for combined moisture and temperature gradients. They predicted changes in moisture content at different depths in tilled and untilled soil. But in none of the earlier models the phenomenon of multiloci moving downward with time has been explained. Therefore for any mechanistic model input data should include the downward movement of the locale of evaporation, the soil temperature wave in tilled and untilled soil, and wind penetration as affected by clod size.

VI. Summary

Literature reviewed in this chapter points to the need to reconsider the generalizations about the time course of evaporation, based on the solutions of the flow equation, which neglected gravity and assumed infinite evaporativity and instantaneous drying of the soil surface. Evidence has become available to show that depending on the soil type, initial wetness, and the E_0 involved, cumulative evaporation from soil under lower E_0 can exceed that under higher E_0. Similarly, the experimental results do not support that the evaporation rate during the falling-rate stage is independent of E_0. This was found true for coarse-textured soil only.

Loosening of the surface soil layer by cultivation alters the time course of evaporation by changing the albedo, surface area exposed to the atmosphere, and penetration of wind into the soil. These changes collectively accelerate the rate of water loss and enhance the development of a dry layer at the surface, which tends to reduce further water loss from soil. Evaporation reduction by shallow tillage, which has been a matter of controversy for the last half-century, depends on the retentivity and transmissivity of the soil, E_0, time (after wetting), and type of tillage, and is time variant. On relatively fine-textured soils under low E_0, the benefit of tillage is greater and continues to increase with time for long periods. On the contrary, on coarse-textured soils under high E_0 there is little benefit from shallow tillage because these conditions are conducive to self-mulching. Between these two extremes there are numerous soil, E_0, and tillage (time and type) combinations where the soil mulch created by shallow tillage would help reduce evaporation losses to varying degrees. Apart from evaporation reduction, tillage causes higher wetness in layers immediately below the depth of tillage.

It is indicated that the locus of evaporation in the tilled soil does not instantaneously shift to the bottom of the tilled layer, as was implied in several earlier efforts to predict tillage effects using surfactant-treated soil mulch (Hanks, 1958; Acharya and Prihar, 1969; Hillel et al., 1975). In fact, upward liquid flow through the junction of tilled and untilled soil continues for variable lengths of time, depending on soil and climate characteristics. Therefore it is postulated that the evaporation plane, constituted by a multilocus downward moving front, is located within the tilled layer for a considerable length of time after tillage. It is suggested that for any mechanistic modeling of evaporation from tilled soil, input data should include the downward movement of the locale of evaporation, the soil temperature wave, and wind penetration.

References

Acharya, C. L., and S. S. Prihar. 1969. Vapour losses through soil mulch at different wind velocities. *Agron. J.* 61:666–668.

Acharya, C. L., S. S. Sandhu, and I. P. Abrol. 1979. Effect of exchangeable sodium on the field drying pattern of soil at two evaporative demands. *Soil Sci.* 127:56–62.

Allmaras, R. R. 1967. Soil water storage as affected by infiltration and evaporation in relation to tillage induced soil structure. In: *Tillage for Greater Crop Production, Conf. Proc.* Amer. Soc. Agric. Eng., St. Joseph, MI.

Allmaras, R. R., W. W. Nelson, and E. A. Hallauer. 1972. Fall vs. spring plowing and related heat balance in western corn belt. Tech. Bull. 283, Minnesota Agric. Exp. Stn., St. Paul, MN.

Archer, J. R., and P. D. Smith. 1972. The relation between bulk density, available water capacity and air capacity of soils. *J. Soil Sci.* 23:475–480.

Arkley, R. J. 1963. Relationships between plant growth and transpiration. *Hilgardia* 34:559–584.

Aujla, T. S., and S. S. Cheema. 1983. Modifying profile water storage through tillage, herbicide, chemical evaporation retardent and straw mulch and its effect on rainfed chickpea (*Cicer arietinum* L.). *Soil Tillage Res.* 3:159–170.

Benoit, G. R., and D. Kirkham. 1963. The effects of soil surface conditions on evaporation of soil water. *Soil Sci. Soc. Amer. Proc.* 27:495–498.

Black, T. A., W. R. Gardner, and G. W. Thurtell. 1969. The prediction of evaporation, drainage and soil water storage for a bare soil. *Soil Sci. Soc. Amer. Proc.* 33:655–660.

Bond, J. J., and W. O. Willis. 1970. Soil water evaporation, first stage of drying as influenced by surface residue and evaporation potential. *Soil Sci. Soc. Amer. Proc.* 34:925–928.

Bowers, S. A., and R. J. Hanks. 1965. Reflection of radiant energy from soils. *Soil Sci.* 100(2):130–138.

Box, J. E., and S. A. Taylor. 1962. Influence of soil bulk density on matric potential. *Soil Sci. Soc. Amer. Proc.* 26:119–123.

Bresler, E., W. D. Kamper, and R. J. Hanks. 1969. Infiltration redistribution and subsequent evaporation of water from soil as affected by wetting rate and hysteresis. *Soil Sci. Soc. Amer. Proc.* 33:832–840.

Budogoveskly, E. I. 1962. *Evaporation of water by soil (heat balance of forest field).* Izd. Akad. Nauk, USSR.

Call, L. E., and M. C. Sewell. 1917. Soil mulch. *J. Amer. Soc. Agron.* 9:49–61.

Cary, J. W., and D. D. Evans (eds.). 1975. Soil crusts. Tech. Bull. 214, Arizona Agric. Exp. Stn., Tuscon, AZ.

Covey, W. 1963. Mathematical study of first stage of drying of moist soil. *Soil Sci. Soc. Amer. Proc.* 27:130–134.

Covey, W., and M. E. Bloodworth. 1966. Evaporativity and second stage of drying of soil. *J. Appl. Meterorol.* 5:364–365.

Crank, J. 1956. *The mathematics of diffusion.* Oxford Univ. Press, London.

de Wit, C. T. 1958. *Transpiration and crop yields.* Ver Landboukd. Onderz. 64, Wageningen, The Netherlands.

Ehlers, W., and R. R. Van Der Ploeg. 1976. Evaporation, drainage and unsaturated hydraulic conductivity of tilled and untilled fallow soil. *Z. Pflanzenernähr. Bodenkd.* 139:373–376.

Eser, C. 1884. Untersuchungen Über den Einfluss der physikalischen und chemischen

Eigenschaften des Bodens auf dessen Verdunstungsvermögen. *Forsch. Geb. Agric.-Phys.* 7:1–124.

Fritton, D. D., D. Kirkham, and R. H. Shaw. 1967. Soil water and chloride redistribution under various evaporation potentials *Soil Sci. Soc. Amer. Proc.* 31:599–603.

Gardner, W. R. 1959. Solutions of flow equation for the drying of soils and other porous media. *Soil Sci. Soc. Am. Proc.* 23:379–382.

Gardner, W. R., and M. Fireman. 1958. Laboratory studies of evaporation from soil columns in the presence of water table. *Soil Sci.* 85:244–249.

Gardner, H. R., and W. R. Gardner. 1969. Relation of water application to evaporation and storage of soil water. *Soil Sci. Soc. Amer. Proc.* 33:192–196.

Gardner, W. R., and D. I. Hillel. 1962. The relation of external evaporation conditions to the drying of soil. *J. Geophys. Res.* 67:4319–4325.

Gardner, W. R., and M. S. Mayhugh. 1958. Solutions and tests of the diffusion equation for the movement of water in soil. *Soil Sci. Soc. Amer. Proc.* 22:197–201.

Gill, K. S., and S. S. Prihar. 1983. Cultivation and evaporativity effects on drying patterns in a sandy loam soil. *Soil Sci.* 135:366–377.

Gill, K. S., and S. S. Prihar. 1988. Seedling emergence at four temperatures from drying out seed zones underlain by wet soil. *Plant Soil* 112:267–272.

Gill, K. S., S. K. Jalota, S. S. Prihar, and T. N. Chaudhary. 1977. Water conservation by soil mulch in relation to soil type, time of tillage, tilth and evaporativity. *J. Ind. Soc. Soil Sci.* 25:360–366.

Hadas, A. 1975. Drying of layered soil columns under nonisothermal conditions. *Soil Sci.* 119:143–148.

Hammel, J. E., R. I. Papendick, and G. S. Campbell. 1981. Fallow tillage effects on evaporation and seed zone water content in a dry summer climate. *Soil Sci. Soc. Amer. J.* 45:1016–1022.

Hanks, R. J. 1958. Vapour transfer in dry soil. *Soil Sci. Soc. Amer. Proc.* 22:372–374.

Hanks, R. J., and W. R. Gardner. 1965. Influence of different diffusivity water content relation on evaporation of water from soils. *Soil Sci. Soc. Amer. Proc.* 29:495–498.

Hanks, R. J., and N. P. Woodruff. 1958. Influence of wind on vapour transfer through soil, gravel and straw mulches. *Soil Sci.* 86:160–164.

Hanks, R. J., H. R. Gardner, and M. L. Fairbourn. 1967. Evaporation of water from soil as influenced by drying with wind or radiation. *Soil Sci. Soc. Amer. Proc.* 31:593–598.

Hanks, R. J., H. R. Gardner, and R. L. Florian. 1969a. Plant growth–evapotranspiration relation for several crops in the central great plains. *Agron. J.* 61:30–34.

Hanks, R. J., A. Klute, and E. Bresler. 1969b. A numerical method for estimating infiltration, redistribution, drainage and evaporation of water from soil. *Water Resour. Res.* 5:1064–1068.

Harrold, L. L., D. B. Peters, F. R. Drebelbis, and J. L. McGuinness. 1959. Transpiration evaluation of corn grown on plastic-covered lysimeters. *Soil Sci. Soc. Amer. Proc.* 23:174–178.

Heinonen R. 1985. *Soil Management and Crop Water Supply*, 4th ed. Dept. Soil Sci., Swedish Univ. Agric. Sci., 575007 Uppasala, Sweden.

Hide, J. C. 1954. Observation on factors influencing the evaporation of soil moisture. *Soil Sci. Soc. Amer. Proc.* 18:234–239.

Hillel, D. I. 1971. *Soil and water: Physical principles and procedures*. Academic Press, New York, London, pp 184, 192–200.

Hillel, D., and A. Hadas. 1972. Isothermal drying of structurally layered soil columns. *Soil Sci.* 113:495–498.

Hillel, D., and H. Talpaz. 1977. Simulation of soil water dynamics in layered soils. *Soil Sci.* 123:54–62.

Hillel, D., C. H. M. Van Bavel, and H. Talpaz. 1975. Dynamic simulation of water storage in fallow soil as affected by mulch of hydrophobic aggregates. *Soil Sci. Soc. Amer. Proc.* 39:826–833.

Holmes, J. W., E. L. Greacen, and G. C. Gurr. 1960. The evaporation of water from bare soils with different tilths. In: *Trans. 7th Int. Cong. Soil Sci.*, Madison, WI Vol. 1. Int. Soc. Soil Sci., Wageningen, The Netherlands, pp. 188–194.

Idso, S. B., R. D. Jackson, and R. J. Reginato. 1975. Estimating evaporation. A technique adaptable to remote sensing. *Science* 189:991–992.

Jacks, G. L., V. V. Brind, and R. Smith. 1955. Mulching. Tech. Commun. 49, Commonwealth Bureau of Soil Sci., Harpenden, U.K.

Jackson, R. D., B. A. Kimball, R. J. Reginato, and F. S. Nakayama. 1973. Diurnal soil water evaporation time depth flux patterns. *Soil Sci. Soc. Amer. Proc.* 37:505–509.

Jackson, R. D., B. A. Kimball, R. J. Reginato, and F. S. Nakayama. 1974. Diurnal soil water evaporation: Comparison of measured and calculated water fluxes. *Soil Sci. Soc. Amer. Proc.* 38:861–866.

Jackson, R. D., S. B. Idso, and R. J. Reginato. 1976. Calculation of evaporation rates during the transition from energy limiting to soil limiting phases using albedo data. *Water Resources Res.* 12:23–26.

Jaggi, I. K., R. K. Gupta, and M. B. Russel. 1978. Evaporation from bare soil. *J. Ind. Soc. Soil Sci.* 26:71–73.

Jalota, S. K. 1984. Drying pattern of bare and tilled silt, loam sandy loam and loamy sand soils as affected by zero time water profiles and evaporativity. Ph.D. thesis, Punjab Agric. Univ. Ludhiana, India.

Jalota, S. K., and S. S. Prihar. 1979. Soil water storage and weed growth as affected by shallow tillage and straw mulching with and without herbicide in bare fallow. *Ind. J. Ecol.* 5:41–48.

Jalota, S. K., and S. S. Prihar. 1985. Evaporation reduction by tillage and redistribution time. In *Abstracts, Int. Cong. Arid Land Research and Development, Tucson, AZ, Oct. 22–25, 1985.*

Jalota, S. K., and S. S. Prihar. 1986. Effects of atmospheric evaporativity, soil type and redistribution time on evaporation from bare soil. *Aust. J. Soil Res.* 24:357–366.

Jalota, S. K., and S. S. Prihar. 1987. Observed and predicted trends from a sandy loam soil under constant and staggered evaporativity. *Aust. J. Soil Res.* 25:243–249.

Jalota, S. K., S. S. Prihar, and K. S. Gill. 1988. Modified square root of time relation to predict evaporation trends from bare soil. *Aust. J. Soil Res.* 26:281–289.

Kijne, J. W. 1973. Evaporation from bare soils pp. 221–226. In: A. Hadas, D. Swartzendruber, P. E. Rijtema, M. Fuchs, and B. Yaron (eds.), *Physical aspects of soil water salt in ecosystems*. Chapman & Hall, London: Springer-Verlag, NY.

King, F. H. 1907. Quoted in L. D. Baver. 1966. *Soil physics*. Asia Publ. House, Bombay, India, p. 281.

Klute, A., F. D. Whisler, and E. J. Scott. 1965. Numerical solution of non-linear diffusion equation for water flow in horizontal column of finite length. *Soil Sci. Soc. Amer. Proc.* 29:353–356.

Laliberte, G. E., and R. H. Brooks. 1967. Hydraulic properties of disturbed soil material affected by porosity. *Soil Sci. Soc. Amer. Proc.* 31:447–454.

Lemon, E. R. 1956. Potentialities for decreasing soil water evaporation loss. *Soil Sci. Soc. Amer. Proc.* 16:160–163.

Linden, D. R. 1982. Predicting tillage effects on evaporation. In: P. W. Unger and D. M. Van Doren, Jr. (eds.), *Predicting tillage effects on soil physical properties and processes*. ASA Sp. Publ. 44, Am. Soc. Agron., Soil Sci. Soc. Amer., Madison, WI.

Minhas, P. S. 1985. Salt and water movement in unsaturated soils as affected by method of water application, surface soil condition and atmospheric evaporativity. Ph.D. thesis, Punjab Agric. Univ., Ludhiana, India.

Minhas, P. S., B. K. Khosla, and S. S. Prihar. 1986. Evaporation and redistribution of salts in a silt loam soil as affected by tillage induced soil mulch. *Soil Tillage Res.* 7:301–313.

Ojeniyi, S. O., and A. R. Dexter. 1979. pp. 273–283. In: R. Lal and D. J. Greenland (eds.), *Soil physical properties and crop production in the Tropics*. Wiley, NY.

Papendick, R. I., M. J. Lindstrom, and V. L. Cocharn. 1973. Soil mulch effects on seedbed temperature and water during fallow in Eastern Washington. *Soil Sci. Soc. Amer. Proc.* 37:307–314.

Pendelton, J. W. 1966. Increasing water use efficiency by crop management. p. 250. In: W. H. Pierre, D. Kirkham, J. Pesek, and R. Shaw (eds.), *Plant environment and efficient water use*. Amer. Soc. Agron., Soil Sci. Sol. Amer., Madison, WI.

Penman, H. L. 1941. Laboratory experiments on evaporation from fallow soil. *J. Agric. Sci.* 31:454–463.

Peters, D. B. 1960. Relative magnitude of evaporation and transpiration. *Agron. J.* 52:536–538.

Peters, D. B., and M. B. Russel. 1959. Relative water losses by evaporation and transpiration in field corn. *Soil Sci. Soc. Amer. Proc.* 23:170–176.

Philip, J. R. 1957. Evaporation and moisture and heat fields in the soil. *J. Meteorol.* 13:354–366.

Prihar, S. S., and S. K. Jalota. 1988. Role of shallow tillage in moisture management. Int. Conf. on Dryland Farming, Amarillo, TX, Aug. 15–19, 1988.

Prihar, S. S., and D. M. Van Doren. 1967. Mode of response of weed free corn to post planting cultivation. *Agron. J.* 59:513–516.

Prihar, S. S., B. Singh, and B. S. Sandhu. 1968. Influence of soil and environment on evaporation losses from mulched and unmulched pots. *J. Res.* (PAU) 5:320–328.

Ramacharlu, P. T. 1957. Rate of evaporation of water in relation to particle distribution in soils. *J. Ind. Soc. Soil Sci.* 5:117–121.

Reynolds, W. D., and G. K. Walker. 1984. Development and validation of a numerical model simulating evaporation from short cores. *Soil Sci. Soc. Amer. J.* 48:960–969.

Rose, C. W. 1968. Evaporation from bare soil under high radiation conditions. pp. 57–66. In: *Trans. 9th Int. Cong. Soil Sci. Adelaide, Australia*. Int. Soc. Soil Sci., Wageningen, The Netherlands.

Rose, C. W. 1969. Water transport in soil by evaporation and infiltration. pp. 171–181. In: P. E. Rijtema and H. Wassink (eds.), *Water in unsaturated zone. Proc. Wageningen Symp.* Wageningen, The Netherlands.

Rowse, H. R. 1975. Simulation of water balance of soil columns and fallow soils. *J. Soil Sci.* 26:337–349.

Sandoval, F. M., and L. C. Benz. 1966. Effect of bare, barley and grass on salinity of a soil cover over a saline water table. *Soil Sci. Soc. Amer. Proc.* 30:392–396.

Scotter, D. R., and P. A. C. Raats. 1969. Dispersion of water vapours in soil due to air turbulance. *Soil Sci.* 108:170–176.

Singh, D., and S. D. Nijhawan. 1944. A critical study of the effect of soil mulch on evaporation of soil moisture. *Ind. J. Agric. Sci.* 14:127–137.

Stanhill, G. 1955. Evaporation of water from soil under field conditions. *Nature* 176(4471):82–83.

Staple, W. J. 1971. Boundary conditions and conductivity used in isothermal model of evaporation from soil. *Soil Sci. Soc. Amer. Proc.* 36:853–855.

Staple, W. J. 1976. Prediction of evaporation from columns of soil during alternate periods of wetting and drying. *Soil Sci. Soc. Amer. J.* 40:756–761.

Tanner, C. B., A. E. Peterson, and J. R. Love. 1960. Radiant energy exchange in a corn field. *Agron. J.* 52:373–379.

Van Bavel, C. H. M. 1966. Potential evaporation: The combination concept and its experimental verification. *Water Resources Res.* 2:455–467.

Van Bavel, C. H. M., and D. I. Hillel. 1976. Calculating potential and actual evaporation from a bare soil surface by simulation of concurrent flow of water and heat. *Agric. Meteorol.* 17:453–476.

Veihmeyer, F. J. 1927. Some factors affecting irrigation requirements. *Hilgardia* 2:125–291.

Wadham, L. A. 1944. The flow of heat through granular materials. *J. Chem. Soc. India.* 63:337–343.

Widstoe, J. A. 1909. Irrigation investigation. Bull. 105, Utah Agric. Exp. Sta., Logan, UT.

Wiegand, C. A., and S. A. Taylor. 1960. The temperature dependence of the drying of soil columns. pp. 169–178. In *Trans. 7th Int. Cong. Soil Sci. Madison,* WI. Vol. 1. Int. Soc. Soil Sci., Wageningen, The Netherlands.

Wiese, A. F., and T. J. Army. 1958. Effect of tillage and chemical weed control on soil moisture storage and losses. *Agron. J.* 50:465–467.

Willis, W. O. 1962. Effect of partial surface covers on evaporation from soil. *Soil Sci. Soc. Amer. Proc.* 26:598–601.

Willis, W. O. and J. J. Bond. 1971. Soil water evaporation: Reduction by simulated tillage. *Soil Sci. Soc. Amer. Proc.* 35:526–528.

Wind, W. O. 1961. Capillary rise and some applications of the theory of moisture in unsaturated soil. Tech. Bull. 22, Inst. Land and Water Mgt. Res., The Netherlands.

Index

MIX
Papier aus verantwortungsvollen Quellen
Paper from responsible sources
FSC® C105338
FSC
www.fsc.org

If you have any concerns about our products,
you can contact us on
ProductSafety@springernature.com

In case Publisher is established outside the EU,
the EU authorized representative is:
Springer Nature Customer Service Center GmbH
Europaplatz 3, 69115 Heidelberg, Germany

Printed by Libri Plureos GmbH
in Hamburg, Germany